T0276933

ACCIDENTAL ASTRONOMY

ALSO BY CHRIS LINTOTT

The Crowd and the Cosmos: Adventures in the Zooniverse

Bang!!: The Complete History of the Universe
with Brian May, Hannah Wakeford, and Patrick Moore

ACCIDENTAL ASTRONOMY

HOW RANDOM DISCOVERIES SHAPE THE SCIENCE OF SPACE

CHRIS LINTOTT

BASIC BOOKS

New York

Basic Books
Hachette Book Group
1290 Avenue of the Americas, New York, NY 10104
www.basicbooks.com
Printed in the United States of America

Originally published in 2024 by Torva in the United Kingdom

First US Edition: June 2024

Published by Basic Books, an imprint of Hachette Book Group, Inc. The Basic Books name and logo is a registered trademark of the Hachette Book Group.

The Hachette Speakers Bureau provides a wide range of authors for speaking events. To find out more, go to hachettespeakersbureau.com or email HachetteSpeakers@hbgusa.com.

Basic books may be purchased in bulk for business, educational, or promotional use. For more information, please contact your local bookseller or the Hachette Book Group Special Markets Department at special.markets@hbgusa.com.

The publisher is not responsible for websites (or their content) that are not owned by the publisher.

Print book interior design by Bart Dawson.

Library of Congress Cataloging-in-Publication Data
Names: Lintott, Chris, author.
Title: Accidental astronomy : how random discoveries shape the science of space / Chris Lintott.
Description: First US edition. | New York, NY : Basic Books, 2024. | "Originally published in 2024 by Torva in the United Kingdom." | Includes bibliographical references and index.
Identifiers: LCCN 2023049966 | ISBN 9781541605411 (hardcover) | ISBN 9781541605428 (ebook)
Subjects: LCSH: Astronomy—Popular works. | Discoveries in science—Popular works.
Classification: LCC QB44.3 .L56 2024 | DDC 520—dc23/eng/20231117
LC record available at https://lccn.loc.gov/2023049966

ISBNs: 9781541605411 (hardcover), 9781541605428 (ebook)

LSC-H

Printing 1, 2024

"It is the stars, / The stars above us, govern our conditions"

(*King Lear*, Act IV, Scene 3);

but perhaps

"The fault, dear Brutus, is not in our stars, / But in ourselves"

(*Julius Caesar*, Act I, Scene 2)

As quoted in "B²FH," paper in *Reviews of Modern Physics*, Vol. 29, No. 4 (October 1957) by Burbidge, Burbidge, Fowler and Hoyle which explained how heavy elements were made in stars.

CONTENTS

Introduction

ACCIDENTAL ASTRONOMY

A little more than five billion years ago, one of the hundred billion or so stars in our Milky Way Galaxy exploded. Stars end their lives once their reserves of fuel, which power the nuclear reactions at their center that make them shine, are exhausted. The immediate result for this massive star was spectacular, producing a supernova, an explosion that briefly, for a matter of days, outshone the rest of the galaxy, but despite its dramatic show the event was not particularly remarkable. A middle-aged galaxy like our Milky Way, an average-sized system engaged in the business of turning gas and dust every so often into stars, perhaps hosts one such supernova once a century.

Like most, this particular supernova will have left behind a dense remnant, either a black hole or else an exotic, dense object called a neutron star. It will also have scattered material throughout its neighborhood, mostly gas newly enriched in heavier elements formed by the nuclear reactions that take place among the debris. It is these reactions, rather than the initial explosion, that produced the light by which the supernova would have shone so brightly so briefly. The shock of the star's sudden end will have borne this material outward, sweeping up and carrying away any surrounding gas to create a void, a bubble 150 light years—a million billion kilometers—across.* At the edge of the bubble, gas was compressed, and the resulting disruption began a process of collapse that, in the course of a few tens of millions of years, produced a new generation of stars.

Among those new stars was one unremarkable yellow dwarf, and five billion years later this star is the Sun that warms the Earth each morning. As it rises sedately and surely each day in the east, as it has done for billions of years, long before any creature existed to gaze up and wonder about its origins, it has come to symbolize for many of our planet's cultures the stability and timelessness of the cosmos. Yet we now know that it owes its existence to the cosmic coincidence of a supernova in just the right place and at just the right time, without which you and I would not be here to contemplate that dawn.**

A similar story of chance could be told of the careening of asteroids in the disc of rubble that surrounded the newly

---

* To deal with cosmic distances, we use the distance that light, the fastest thing in the Universe, can travel. On this scale, the Sun is eight light minutes away, the nearest star four light years away, and our nearest big galaxy, Andromeda, 2.2 million light years distant. Light travels roughly a foot per nanosecond, which means you see your feet not as they are now, but as they were five or six-billionths of a second ago.

** Usually, I contemplate it by hitting snooze and hiding under the duvet, but perhaps you are a morning person.

formed Sun and within which Earth formed, or the collision of a proto-planet the size of Mars with the newly formed Earth, a cataclysm that produced the Moon, whose gravitational pull causes the tides that washed our distant fishy ancestors onto unfamiliar shores. A little later, an asteroid coincidentally happened, unfortunately for the dinosaurs, to be in the same place at the very same time as the Earth, so greatly improving the chances of the small furry mammals that survived the impact and went on to evolve, before developing astronomy, gin, and the other trappings of civilization a little while later.

Further back in cosmic history, five and a half billion years or so ago, a small galaxy approached the growing Milky Way. This encounter may have triggered the formation of the star that became the supernova I started with. A series of such collisions are likely responsible for our galaxy's structure, right down to its beautiful spiral arms, as well as our Sun's existence among its neighboring stars. When we look up at the night sky, where once we might have imagined stars divinely arranged in their unchanging patterns, moving in perfect order, science tells us we should see only an accidental Universe, arranged by chance.

Trying to understand this feels less like an exercise in cerebral theoretical physics and more like reading history. There is no equation scribbled on a blackboard that can tell you why the Milky Way, Sun, and Earth are precisely the way they are, or why this planet in particular has produced a species capable of trying to tell their stories in the first place. We exist as the result of a chain of countless accidents, just as my happening to be sitting at this desk on a January day in Oxford is the result of historical happenstance on scales large and small.

This isn't to say that science is powerless in such circumstances. There are equations, theories, and physical laws, each derived from physics, chemistry, and their allied trades, that can tell us what processes are at work in the cosmos and what the odds of various outcomes are. Ask not about why a particular galaxy called the Milky Way sits in this particular part of the cosmos, but instead about the chances of galaxies that are like the Milky Way existing, and we are in business. The history of the Earth itself is difficult to pin down, but trying to work out how many of the billions of planets we now think pepper the cosmos are similar to the Earth is a more tractable problem. It is a major goal of modern astronomy, with new telescopes and much brainpower directed to this burning question. And while I can't predict your fate, I can tell you that the odds of an asteroid the size of the one that did in the dinosaurs hitting us next year are around one in fifty million, about the same as the odds that you will die in a plane crash at some point.

These, though, are statements about probabilities. Certainties come only when we look at simple systems, and usually only on short timescales. Of the eight hundred thousand or so known asteroids, the most threatening to us is Bennu, a pile of rubble a few hundred meters across that was discovered in 1999 by a telescope operated by the US Air Force. Like everything in the Solar System, Bennu travels a path in orbit around the Sun under the influence of gravity, and I can use the equations discovered by Isaac Newton to predict how it will move. (I might, if I wanted to be especially accurate, upgrade to Einstein's theory of relativity, which refined Newton's work, but in this case the difference will be extremely small. The *Apollo* astronauts navigated to the

Moon and back using Newtonian calculations, and Isaac will do fine for asteroids.) Our understanding of gravity has allowed us to follow Bennu in the sky since its discovery, and it allowed me to use a friend's telescope to take a look at it just the other night, the computerized drive whirring as it moved to the predicted coordinates to show us the asteroid, visible as a point of light seen against a background sprinkling of stars. It has even been good enough to take the OSIRIS-REx spacecraft from Florida on a journey of millions of kilometers to visit the asteroid and return a sample of it back to us here on Earth, the better to understand how we might protect ourselves against future impacts.

Our ability to send spacecraft like OSIRIS-REx zipping about the Solar System, visiting asteroids whenever we choose, might suggest that we are able to predict Bennu's path perfectly, looking as far into the future as we need to reassure ourselves that it poses no threat, but this is simply not the case. We would do a better job of prediction in a solar system that consisted of only the Sun and Bennu, but here in our crowded neighborhood the gravitational pull of each of the planets, added to that of our central star, has a small but tangible effect on Bennu's motion. The position of giant Jupiter matters in the long term, as does the effect of the asteroid's occasional close encounters with the Earth. There are also forces other than gravity at work. The Sun's light exerts a gentle, almost infinitesimal pressure on the asteroid, capable over decades of appreciably altering its position from where it would be if only gravity mattered.

The best we can do is to tell you that the odds of an impact in the twenty-first century, or even during the first half of the twenty-second, are very low indeed, though at the time of writing there's a 1 in 1,750 chance of an impact that would occur in the year 2178, or on several possible dates in the years that follow. If you are both of a particularly pessimistic nature and really into long-term planning, you might make sure to be somewhere with a strong roof on September 24, 2182, lest Bennu come crashing down. As we get closer to any specific encounter, our ability to predict what exactly will happen, and to distinguish incoming catastrophe from a close shave, will improve. It is most likely that the chances of an impact on any particular date will decrease as time goes by, but maybe they won't. We can't tell for sure, and no amount of staring at Bennu or hard work on new theories can help us.

On a grander scale I can tell you about the processes that dictate how, and to some extent why, galaxies formed, emerging from what was a bland, uniform soup of matter that existed shortly after the Big Bang to create the great cosmic web of today, its superclusters containing thousands of large systems linked by long filaments surrounding cosmic voids. Indeed, with the slightest encouragement I will do so, but I can't describe in detail why the Milky Way Galaxy turned out to be a spiral. In order to understand the physics that governs the evolution of the cosmos, astronomers make simulations that take years of work to program and that require some of the most powerful supercomputers in the world to run for months, producing virtual universes we can explore. What they can't do is produce an exact replica of our Universe.

Knowing this, looking up at the night sky or at an image like the Hubble Deep Field, which in a single shot shows thousands of galaxies each the equal of the Milky Way, can make me feel dizzy. Our place in this vast and awesome cosmos is so precarious. We could very easily not have been here to experience it: any one of a number of chances could well have fallen the other way. Simultaneously, such thoughts fill me with glee. I am almost unable to believe my luck that, despite overwhelming odds, I get to see such beauty and grandeur, even while experiencing a touch of terror at being lost in a vast and accidental Universe.

This twin perspective, of both awe and joy, is a feature of thinking about astronomy even when we stay closer to home. When the Curiosity rover, which spends its days exploring Gale Crater, an ancient lake bed on Mars, sends back new pictures to Earth, they are almost instantaneously posted online. If you happen to click refresh at the right time, you can be the first to see its holiday snaps, returned from an alien world. Most are as mundane as an ill-chosen postcard of some metropolis's bus station, stuck on the bottom of a neglected postcard rack at a newsstand: selfies of calibration targets, such as the miniature sundial placed on the rover deck, are common, and there are many close-up shots of rocks that the rover's robot arm and drill may (but probably will not) be asked to prod and poke. But just occasionally, when Curiosity and its team of rover drivers are planning the next stage of the exploration of that ancient landscape, its cameras look outward and provide views of simply magnificent splendor.

A recent view that popped up on my screen showed the entrance to a shallow valley, its rock-strewn floor stretching between rounded mesas. One of these looked like a Martian version of the famous Half Dome in California's Yosemite, and only a blink is needed to add humanity to the scene, imagining the tents of climbers perched precariously halfway up. There may be no campers, yet, and no vegetation to provide a splash of green to break up the red rocks, but there is surprising variation in color. Lighter, pinker rocks contrast with dark, almost brown stripes that run through the landscape. Dust, hanging in the atmosphere, gives the sky a hazy feel, the soft light imparting a twilight sense to the scene.

It is easy to imagine standing in that very spot. Perhaps it is because the mast of the rover, which holds the camera, is about two meters above the ground, and so the images from Curiosity and many of its fellow robot explorers offer a very human perspective on the planet. Looking at them sets me thinking of the time I've spent hillwalking,* and makes me want to start

---

* Admittedly, a large proportion of this was in the company of the Cambridge University Hillwalking Club, who met at the Castle Inn halfway up the only thing that could even vaguely be called a hill in that notoriously flat Fenland town.

planning a route for a wander through this landscape. Friends who climb have told me of similar experiences, picking out their pitches for the first ascent of these hills, while still others, facile with a paintbrush, see the terrain with artist's eyes. Many have been inspired to work with the images themselves. A community of digital photographers, image processors, and graphics experts spend their spare time producing cleaned-up, often gorgeous renderings of what the rovers are sending back from Mars, even collaborating with the engineers and scientists who built and operate Curiosity to get better pictures and help plan further extraterrestrial photography.

Visiting Mars through their efforts can feel as familiar as flipping through the travel section in the Sunday papers, pausing over artful images of places that I have never been to, but can

The Curiosity Mars rover looks into Gediz Vallis, a valley cut into the side of Mount Sharp. The scattered rocks on the valley floor offer us the chance to stumble across younger material from high on the slopes of the mountain. *Credit: NASA/JPL-Caltech/MSSS/Kevin M. Gill*

dream of one day visiting. With images from Curiosity in hand, it seems as likely that I might wander Gale Crater or hike Mars's version of the Grand Canyon, the Valles Marineris, as I might one day take the advice of some journalist and find myself kicking back and catching rays at a resort in Bali.* Though Mars is millions of kilometers away, and it remains far beyond our capabilities to send anyone there, it can seem familiar.

One of my favorite Curiosity images was taken on January 31, 2014, about an hour and a half after sunset. There is a darkened horizon, craggy and distant, just illuminated by the fading light of a hazy sky. Twenty degrees or so above the horizon is a bright evening star, a scene as familiar as can be, reminiscent of all of the times that I've stepped through the back door, looked up, and seen Venus, Jupiter, or Mars itself emerge as the first "star" visible after sunset, marking the start of another clear night.

Look more closely and you'll see not one star in the Martian sky but two, close together yet still distinguishable. One, with a blueish tinge, is brighter than its whiter companion. Hanging in Mars's twilight, this is the Earth and the Moon, points of light containing all of humanity and our hopes and dreams, reduced to just two stars in another world's sky. The effect of this realization is enough to knock me off balance. The familiar landscape I imagined walking in suddenly becomes impossibly alien, an otherworldly place that seems distinctly out of reach.

Both sets of feelings, of a planet just next door I can imagine visiting and a world that seems foreign and remote, hit me at

---

* I'm not good at doing nothing, which is a distinct problem with a hypothetical vacation in Bali, for one thing. At least on Mars I could explore, though the lack of a breakfast buffet is a definite negative.

once when I look at those images of Mars. It's the same duality as comes when looking at images of distant galaxies, flipping between being at home in the Universe and cosmically alienated, secure in our Universe of ordered physics and contemplating an infinity full of accidents. We have learned, ever since the Hubble Space Telescope first flew, that if you point a telescope at even the emptiest patch of sky and leave its camera shutter open long enough, the camera's field of view will slowly become speckled with fuzzy galaxies, some no more than faint points, whose light has taken billions of years to travel through the Universe before, by accident, encountering our camera. Such images have told us that there are more stars out there in the observable Universe than there are grains of sand on Earth.* We are adrift in this cosmic vastness, and it is hard not to feel that our planet and all its inhabitants are surely insignificant in the face of such immensity. As Douglas Adams memorably commented in his *Hitchhiker's Guide to the Galaxy*, space is big, but it can make us feel small.

This feeling doesn't go away as you learn more; you can't think yourself out of it. No amount of poring through books on cosmology, no number of lectures from eminent physicists, no number of hours spent watching Carl Sagan on TV or sitting under waterfalls cross-legged will lead you to an intuitive understanding of what it means to live on a planet that could be one of billions upon billions scattered through the darkness. As an astronomer, I have no more profound insight on what it means to live in our Universe than you do, though I suspect that sheer

---

*   Probably; it turns out we know the number of stars in the Universe more accurately than we know the number of grains of sand on the Earth. Pull your collective finger out, geologists.

repetition has meant that I've gotten better at not worrying too much about such things when confronted with them on a rainy Monday when there's work to be done. A sense of awe and wonder is simply the only appropriate response to finding ourselves in this frankly enormous Universe, and it is, I think, an inescapable part of our cosmic experience.

Maybe we can take some solace in the fact that, amid such vastness, we may stand out. It seems likely that the Earth has millions of twin planets out there among the stars; a recent estimate suggests that perhaps at least a billion Earth-like worlds[*] exist in the Milky Way alone. Yet, despite our search for alien life and the dreams of several centuries of science-fiction authors and fans, our planet is still the only one where we know that life has reached the level of sophistication that it has around here. Olaf Stapledon, the philosopher who wrote some of the strangest and most inspiring science fiction of the twentieth century,[**] said in his history of the next few hundred billion years, *Last and First Men*:

> This is the goal of all living, that the cosmos may be known, and admired, and that it may be crowned with further beauties. Nowhere, and at no time, so far as we can tell, at least within our own galaxy, has the adventure reached further than in ourselves. And in us, what has been achieved is but a minute beginning. But it is a real beginning.

---

[*]   Technically, worlds that, if they shared Earth's size, mass, and atmosphere, might be the right temperature to have liquid water on their surface.

[**] Do read *Last and First Men*, but skip the first eighty or so pages.

Imagining ourselves as the pinnacle of evolution in this way lets us look out proudly and with something of a proprietary air at the Universe, as we bravely undertake our cosmic task of investigating our surroundings on its behalf. This is a pleasant trick for astronomers, both amateur and, in particular, professional, to play. As my colleague, exoplanet expert Jayne Birkby, said to me recently,* astronomers must have the ego to comprehend that the Universe is impossibly, imaginably vast, and yet still think it will make a difference whether they understand it or not.

Yet the idea of a lonely Universe in which we—a species whose loudest, barbaric yawp of a broadcast into the cosmos is still, as we'll discover later, an advertisement for Doritos, and who here at home can produce acts of horrible cruelty as easily as we can great leaps of artistic flair or scientific understanding—are as good as it gets is hard to stomach. Equally dizzying is the opposite prospect, that every other planet from Proxima Centauri b on out has a conveniently humanoid alien race, each also looking to the stars and wondering about their neighbors in a shared dream. Both a Universe that is empty of anyone except us and one teeming with life seem impossible to imagine. Are we a cosmic accident, or the inevitable result of the way that physics and chemistry sculpt matter in a Universe like ours? We don't know.

Whatever the rules of the game may be, the rolling of the cosmic dice has placed us here, and luck also plays a role in our attempts to understand our circumstances. If we had been born on a planet that orbited a star at the heart of one of the great globular clusters that surround the main bulk of the galaxy,

---

* In the Lamb & Flag, now reopened after a tough few years and gloriously once again the home of Oxford scientific discussion.

astronomy would be very different. Any intelligent being living in these grand concentrations of hundreds of thousands of stars would see a spectacular night sky, glowing with the light of countless objects brighter than Venus is in ours. But the brilliance of their neighbors would surely make it hard to imagine a Universe beyond. The cosmology of such creatures would be limited just to their natal cluster, and the grand view we have of the cosmic web of galaxies would be denied to them.* As I note later, any creatures living within the ice-roofed oceans of a world like Jupiter's large moon Ganymede may not conceive of anything outside their small, watery bubble. Our vantage point is, luckily, not nearly so constraining.

We live, also, at a special time in the Universe's history. In the first billion years or so after the Big Bang, there was hydrogen and helium, but before the first stars and the first supernovae there were none of the heavy elements that make up you, me, or the world around us.** Life could therefore only get started after the first cycle of star formation and destruction had taken place, seeding its surroundings with materials for the planet building that could accompany the next, second, generation of stars, but such propitious circumstances will not last.

The bad news is that our Universe is now past its peak. Star formation was at its most vigorous several billion years ago, and there are more stars dying each year than are being born. Our cosmological observations suggest that the expansion of the Universe that started at the Big Bang is speeding up, admittedly for

---

* Admittedly, they're going to have a shock when they get to radio astronomy.
** Astronomers call these "metals," which drives chemists nuts. To us, oxygen, nitrogen, and carbon are all metals.

reasons we don't understand, which leads us to a bleak future that—worse—turns out to be rather boring. As the last stars die, in roughly twenty billion years' time, our dynamic and beautiful Universe will contain nothing but a sea of particles and light, expanding forever but in which nothing ever happens. Faced with such a future, we should maybe count ourselves lucky to be living at a time when we have a cosmos filled with galaxies and spectacular sights to observe, rather than finding ourselves on a planet in orbit around one of the last few remaining, isolated stars.

What we get to see of the Universe during our lifetimes is, of course, also a matter of chance. Light from the last supernova observed in the Milky Way Galaxy reached Earth in the year 1604, just before the invention of the telescope. Though it was observed by astronomers around the world, including the great Johannes Kepler, for whom it is usually named, it was little understood. Just a century or two later, such an event would have galvanized a worldwide observation campaign and illustrated the importance of such explosions. A bright supernova in the nineteenth century might have inspired new ideas about how stars work, pushing forward astronomy by a century or so. Instead, we're still waiting on the next nearby one, telescopes and instruments at the ready, replete with questions that can only be answered by studying the death of a star up close.

While we wait, we make do with observing more distant explosions. The nearest supernova seen in the twentieth century, SN 1987A, took place in the Large Magellanic Cloud, one of the Milky Way's satellite galaxies; its remnant, a glowing ring of shocked gas expanding outward from the site of the explosion, can still be monitored thirty-five years later by large telescopes here on Earth. 1987A was close enough for a handful of neutrinos, tiny particles produced in the explosion, to be accidently detected by experiments particle physicists had constructed to study neutrinos from the Sun. These results were so exciting that dedicated experiments have been built to try to capture neutrinos from the Milky Way's next supernova.

The most advanced of these is situated in Gran Sasso, an amazing particle physics laboratory in, or rather under, the mountains to the east of Rome. You reach the lab's doors by taking a sudden and unexpected turn off the road that carries a motorway through the mountain and punching a code into a panel that makes a large door, perfect for any spy movie villain, slowly trundle out of your way. The laboratory is here because the rock above it shields the experiments it contains from cosmic rays. These high-energy particles that rain down from space normally pass harmlessly through you, but they constitute a bright background that swamps detectors looking for other things, like MaCHO, a German experiment that sits waiting for a convenient supernova.

When I visited in 2017, the experiment had been running for more than a decade without success and needed an upgrade. Yet the team I spoke to, hard-headed scientists and engineers all, were wary about shutting down their detector for the eighteen months or so that such improvements would take, feeling that Murphy's Law would inevitably ensure that a supernova would occur in the crucial period.

Looking deeper into the Universe, chance affects us in a different way. I have friends who spend their lives studying the

Andromeda Galaxy, the nearest large system to our own Milky Way, and one that will, in about four or five billion years' time, collide with us. It is a fascinating system that seems to have undergone some sort of merger already, producing a bright ring of star formation that is spreading out from the center of the galaxy. It is, like the Milky Way, a disc galaxy, a thin pancake of a thing which, maddeningly, is turned almost edge on to us. We have only an oblique view, by random chance, of this fascinating and important object; we cannot, unlike scientists studying objects in a laboratory, turn Andromeda over for a closer look.

Galaxies and astronomical objects evolve and interact, driven by processes that unfold over millions and billions of years. We see only snapshots presented to us by the Universe, moments frozen in time, and from them must try to tell the cosmic story. Faced with such restrictions, our best bet is to use every ounce of ingenuity and guile that we have to collect as much information—as many postcards—as we can, hoarding this precious evidence just as archaeologists cling to each pot shard or ornament from the site of a dig.

Sometimes, I should say, it is different. In our quest to understand the history of the Universe, things do occasionally turn out exactly as predicted, with all our observations lined up to confirm some theory or other, but those days are few and far between, and they aren't the really fun ones. We astronomers like being surprised, to wallow for the moment in the sense that there is more to understand. It's a different feeling, utterly, from the way that science and scientific progress are often portrayed on screen or in print, where you're likely to hear stories about singularly clever people who have been blessed, with some clap of thunder, with a dose of cosmic truth before spending their careers trying to prove themselves right. The astronomy I know and love is more likely to involve a bunch of people staring at a screen and looking confused than to feature someone running down a corridor shouting "Eureka."

I mentioned the idea that science was often driven not by deliberate moves but by stumbling accidents to Meg Urry, professor of physics and astronomy at Yale, former president of the American Astronomical Society, and more importantly, one of my favorite people, at a recent conference.* Meg is wise, kind, thoughtful, and smart enough to have been the person to work out how black holes at the center of galaxies behave. As you'd expect from someone so distinguished, she agreed with me completely, going so far as to claim that she couldn't think of a single major discovery that had been planned in advance. We did consider, as a possible exception, the surprising finding made a few decades ago that the expansion of the Universe that began in the Big Bang is speeding up, not slowing down, under the influence of some unknown force. Two sets of astronomers, each carefully measuring the brightness of special supernovae, made the same discovery at the same time in what could be cited as a perfect example of normal, preplanned science, until you learn that both teams had set out to measure how fast the universal expansion was slowing down. Very few people had seriously considered that anything other than slowing under the influence of gravity was possible, and the results were certainly a complete surprise to those doing the work. The cause of this unexpected acceleration,

---

* We have a tradition of having precisely one excellent gin and tonic together at each meeting we both attend.

perhaps a previously unsuspected kind of force that accounts for as much as 70 percent of the energy in today's Universe, remains the greatest mystery in cosmology, and it was found by astronomers looking for exactly the opposite effect.

The stories that follow are my favorite examples of the times when astronomers have stumbled on new truths about the cosmos, either through unexpected discoveries or by suddenly finding new ways to explore. Some of these accidental findings now form the bedrock of our modern understanding of the cosmos, while others may, in the long run, remain mere curiosities. They cover everything from the sudden arrival of a meteorite on a Sunday evening in the English countryside to the oldest light in the cosmos, from an unusual icy moon orbiting Saturn to a misbehaving star whose fluctuating brightness suggested to some that it might be home to an alien civilization with a taste for grand projects. We start with the most ambitious of astronomical quests, the story of the search for aliens, for extraterrestrial intelligence, for a serendipitous discovery that could change the world, but one that has yet to bear fruit.

These stories help me resolve how I feel about my place in the cosmos, and by telling them to you I hope that I can show you how we are trying to understand it. I hope to encourage you to spend a little bit of time contemplating the cosmos too. I want to share how science actually operates, with all the excitement and confusion and randomness that it entails. If I, sitting on my little planet looking up, feel alienated by the inevitable element of chance that is inherent in forming our vast and unfeeling cosmos, then it is because I've forgotten that almost everything we know about it comes from that same rolling of the dice. We are products of randomness, but so too is the arrival of a photon of light from a distant supernova, bearing news about a galaxy far away and the Universe it left long ago.

For as long as our species is capable of it, our position in the Universe leaves us with a responsibility to awe, to borrow a phrase from the astronomer and poet Rebecca Elson: a duty as a species, and as individuals, to make the most of what the Universe provides and to use it to try to understand it. As we look out at the night sky, with the latest telescopes or even with just the naked eye, it's obvious that there are more surprises to come out there in the vastness of the Universe, arrayed among the familiar patterns of the stars.

Chapter 1

IS IT ALIENS?

Most astronomers will have a story of a time where they were caught unawares, a moment when their heart beat slightly faster for just a few seconds wondering what something they had spotted up there in the sky really was. My own close encounter came on a summer's evening, standing on a beach on the south coast of England, looking out at a clear western horizon just after the Sun had sunk below the water. Brilliant Venus was already there, shining brightly in the twilight and twinkling madly, flashing through every color available as its light wobbled in the air heated by the day's warmth, rising off the beach.

I was hoping to catch sight of Mercury, the most elusive of the five planets visible without binoculars or a telescope.* Though it appears as nothing more than a moderately bright

---

*   Six, including Earth.

star to the naked eye, and, being smaller than the Moon and never especially close to us, offers nothing beyond the sight of a plain disc to the telescopic observer, there is always something satisfying in seeing Mercury. I think it's partly the remnant of the kid who collected stamps, still inside my head and wanting to tick each of the Solar System's denizens off my I Spy list; but maybe it's also the satisfaction of seeing the clockwork of the Solar System continue to operate as planets appear in turn and on cue. Whichever it is, making the effort to see our innermost planetary neighbor when it appears in either its morning or, more likely, evening apparition is something I've always done.

I especially look out for it when it is close to the Moon or another planet; without such a signpost, Mercury never really stands out. On this particular evening it was supposed to be a few degrees below Venus, and so I was pleased when the second object I spotted in the darkening sky was a fainter, star-like point of light in what seemed to be the right part of the sky. What was unexpected was that this new object brightened suddenly until it nearly matched Venus, which should have been easily outshining everything else. I was even more astonished when the interloper began to move, first traveling upward away from the horizon, but then taking what appeared to be a sharp right turn. Nothing in the sky, apart from the flashing lights of airplanes, does this naturally.

I knew that the International Space Station (ISS) was far from the sky over the UK, and in any case what I was seeing wasn't behaving like a satellite; they move slowly and consistently in a single direction. In contrast, whatever this was started to move erratically, carving switchbacks in the sky as it rose higher above the horizon. It seemed for all the world to be under intelligent control, a genuine Unidentified Flying Object (UFO), and thoughts of science-fiction aliens, flying saucers, and little gray men muttering "take-me-to-your-leader" did begin to nag at the edge of my consciousness, even while most of my brain was busy running through other, more likely and more prosaic explanations.

Had it blinked out or turned tail back toward the horizon, I might still be wondering what it was. Fortunately, it soon passed nearly overhead, revealing itself to be nothing more than a Chinese lantern floating down the coast, a paper balloon suspended and given lift by the flame of a small candle hanging underneath. What I had taken to be a distant light was in fact much closer than I had anticipated, and what had appeared to be wild changes of direction were nothing more than the thing drifting gently in the light breeze. Given a modicum of confusion and a sprinkle of the unfamiliar, it wasn't hard to find myself half convinced I was seeing intelligence in the sky.

As I said, everyone who looks at the sky has this kind of story, and even casual observers, people whose eyes don't automatically flick upward when stepping outside after dark, occasionally notice what's happening above their heads. There's a long tradition of misidentifying bright lights in the sky. Whenever Venus reaches its greatest western elongation, the part of its orbit where it is most brilliant in Earth's evening sky, I know my inbox will fill with queries from well-intentioned correspondents wondering if the bright thing they've seen might just happen to be the neighbors from Proxima Centauri popping by to borrow a cup of sugar or, as per tradition in 1950s UFO reports, to steal a cow from the prairies of the American Midwest. Satellites fainter than the ISS, moving among the stars, can startle, and I know several people who have been visibly shaken by spotting the US Navy's NOSS surveillance satellites, occasionally visible to the naked eye and arranged so they fly across the sky in a triangular formation that looks anything but natural.

Though they might be "unidentified" to a particular observer, such UFOs are usually readily explained. The large telescopes with which astronomers survey the sky these days, with powerful software dedicated to tracking things like asteroids which move among the stars, do not produce large catalogs of mysterious objects swooping by, and the fact that almost all of us now carry a decent camera attached to our phones has not produced a vast new crop of convincing images of visiting spacecraft. It is perhaps not a coincidence that as our ability to survey the sky and keep an eye out for truly unexpected visitors has improved, the focus of those who argue that there really are alien ships in the skies above Earth has tended to shift from astronomical mysteries to a new set of unusual objects they find in blurry videos, often recorded by high-altitude pilots. These days, I'm much more likely to encounter someone who believes that such videos show evidence of extraterrestrial activity than I am someone who thinks they've spotted spacecraft in the night sky.

The occasional object is picked up, though. Take J002E3, for example, discovered in 2002 by prolific asteroid-hunting amateur astronomer Bill Yeung, which was quickly determined to be in an unusual orbit around the Earth, in a region where interactions with the gravity of the Earth, Moon, or Sun should have ejected it before too long. Did the fact it was there mean that it had been placed in this orbit deliberately, perhaps on a reconnaissance mission?

Studies showed that J002E3 should have spent much of its time orbiting the Sun, changing orbit to circle the Earth just before it was discovered. This was exactly the sort of behavior that you'd expect from an alien surveillance craft, swinging by from time to time for a mission to study the Earth and its inhabitants, perhaps checking every few decades to see if we've worked out how to take care of our planet. Nothing moving like it had ever been seen before. Further investigation, including the discovery of the signature of titanium dioxide in its spectrum, showed that it was not, in fact, the work of aliens. J002E3 seems to be the upper stage of the Saturn V rocket that lifted the *Apollo 12* spacecraft and its astronauts toward the Moon, still hanging around the Earth more than thirty years after the completion of its mission. No one had bothered to track it as it wandered deep space, so when it showed up near the Earth again it was cataloged as something new.

Aliens have presented themselves as the explanation for more complex phenomena than just moving lights in the sky. Everything from the Ashen Light, a nighttime glow observed in the nineteenth century on the dark side of Venus, to the hexagonal feature that appears in images of Saturn's north pole has been blamed on alien activity, though in both cases natural explanations are readily available. The great eighteenth-century observer William Herschel, discoverer of Uranus,* even believed that his

---

* He wanted to call it George after the king, which would have made up in avoiding sniggering what it loses in classical grandeur.

observations showed that the Sun was likely to be inhabited. He interpreted the features that he could see on the solar surface as those of a mountainous landscape, which made the Sun some sort of planet. To Herschel and his contemporaries, planets were quite simply and inevitably homes for intelligent beings, and he wrote that evidence of them wandering his solar landscape would surely be easy to find.

We seem, as a species, predisposed to see the work of intelligence everywhere; I vividly remember one of the pictures sent back by the Spirit Mars rover in 2007, on its mission to understand the geology of the red planet, which seemed to show the blurred image of a crouching gray figure, clearly distinguished from the red soil behind it. Though the scale is wrong, the resemblance to faded images that purport to show Bigfoot in North American forests is clear, and the idea of a tiny Sasquatch on Mars is certainly appealing. It is, however, just a rock.* The infamous face on Mars, which stared back at us, Sphinx-like, in images of a region called Cydonia taken in the 1970s by one of the Viking orbiting spacecraft, was shown in later images to be nothing more than a hill, the deep shadows of which had created the impression of eyes and a nose.

Galle Crater, a little farther north and west than Cydonia, contains within it a curved mountain range with a couple of outlier features that make it look like a classic "smiley" emoji, beloved of ravers in the 1980s. It should perhaps be the spiritual

---

* It may even be an interesting rock! We'll never know, as Spirit did not investigate further, apparent similarity to mythical beasts of the American Northwest not being a priority in planning for those directing the rover's mission of exploring the red planet's watery past.

home of the phenomenon of pareidolia, our tendency to see meaningful images, such as faces, in things wherever possible. It is hard to look at the unfamiliar and not see our own image, a particular hazard for those who are planning to search deliberately for aliens. It would be very easy to expect other intelligent beings to have the same desires, technology, and abilities as ourselves, and thus to reveal themselves in much the same way as we do, whereas that might prove to be very far from the truth.

Flagstaff, Arizona, is an old railway town, a staging point for people exploring the wilderness or, these days, a common brief stop for those en route to the spectacular Grand Canyon. When I first went there as a student in 2001, touring the US on a budget rail pass, I remember spending a sleepless night in a motel next to the tracks, kept awake by the whistles of a passing express or the apparently endless shunting of the occasional mile-long goods train. A little bit of excitement played a role in my insomnia too, as I lay awake looking forward to the next day's visit to Mars Hill, one of the iconic sites of astronomical history.

Located just out of town and accessible via a track that winds through patchy forest, the hill is home to the Lowell Observatory, founded in 1894 and named after its first benefactor, Percival Lowell. Using his family fortune to endow his observatory with state-of-the-art telescopes, one of which would later be used to discover Pluto,* Lowell intended to follow up on his fascination

---

* Its name contains a subtle tribute to PL—Percival Lowell.

with Mars, and particularly with his interest in observations that had been made by an Italian observer named Giovanni Schiaparelli in 1877.

Every few years, there's a period when Mars is close to the Earth. Its disc appears larger, allowing details on the surface to be distinguished. Still, looking at Mars through a telescope, even at these favorable times, is often a frustrating experience. The planet's color, a product of its rusty deserts, is easy to see, and the bright white polar caps that grow and shrink as the seasons pass are also obvious with even modest optical aid. Beyond that, if you're lucky and the skies are clear enough, you might catch sight of a pattern of darker markings. These occasionally disappear from view, hidden by the global dust storms that can engulf the planet, before reappearing in the same place once the weather clears. This ability to regenerate led astronomers in the early twentieth century to suggest that the dark areas might be patches of vegetation, growing back bravely through the dust. They are tough to see even with large telescopes, and through my small back-garden reflector I need to use averted vision—the astronomer's trick of looking out of the corner of one's eye, the better to see faint features—to have any chance at all.

Using one of the largest telescopes in the world, Schiaparelli had spent longer than anyone else looking at Mars, mapping the faint and barely determinable markings he found he could see during those special moments every observer learns to wait for, when the Earth's atmosphere suddenly goes still and the view sharpens. He was the first to discern, among the dark patches and dusty soil, a series of linear features that seemed to stretch across the planet's deserts. He called these *canali*—Italian for "channels"—and over the course of the next few years produced ever-more-detailed maps that showed the network of straight lines expanding, with many of the channels doubling up and even, occasionally, tripling.

In his publications he was careful to avoid offering an explanation for what he was seeing, and he seems not to have been sure whether he was observing a natural phenomenon or something stranger. Lowell had no such qualms. These features to him weren't mere channels but *canals*, the product of intelligence, and of a vast civilization clearly capable of grand works. The Martians were engaged in a vast terraforming effort, bringing water from those polar ice caps to the dry equatorial deserts to irrigate crops or to water cities. Using the telescopes at his new observatory, Lowell's astronomers made a succession of increasingly detailed maps of the planet, showing the sophistication of the growing canal network.

Observers elsewhere in the world were less convinced. Attendees at a discussion at an early meeting of the British Astronomical Association in 1890, for example, were baffled that they could not see any trace of the features reported by an American guest. Of course, the instruments on Mars Hill, particularly the twenty-four-inch refractor that sits there to this day, were some of the most sophisticated and expensive in the world. So was it really that much of a surprise that the Americans could see things that the British, among others, could not?

Life was commonly believed to exist on Mars. In 1900, the Académie des sciences could still support the establishment of a prize to be awarded to the first to communicate with beings on another celestial body, expecting it to be won quickly. Mars was officially excluded, as communicating with the Martians was thought to be too easy.*

When I visited Flagstaff, Mars happened to be visible in the evening sky. The twenty-four-inch telescope was part of the tour, and it was pointed at the planet. I stepped up to the eyepiece to see what Lowell's observers had spent so much time staring at. Mars looked stunning in the enormous instrument, giving me easily the best view of it I can remember, and the details on the disc were more obvious than I had ever seen them, thanks to the ability of the telescope's large lenses to collect plenty of light. Even with such a good view, there was no sign at all of the canals, I'm sorry to say.

Not that I really expected any different. The wondrous waterways of the red planet had long since been revealed to be an illusion, set aside as still larger telescopes failed to reproduce what had been seen from Mars Hill. As photography replaced sketching as the best way to record and share observations, the canals simply failed to show up. A trick of the eye or just the product of overactive imagination? There has been no serious allegation of fraud to explain how experienced scientists, over the course of more than a decade, recorded what is simply not there. It seems to be the case that Lowell's observers, coming up against the limit of what could be known about a planet tens of millions of kilometers away, had simply seen what they expected

---

* Perhaps wanting to get rid of the responsibility, the academy eventually spent what was left of the prize money after inflation on medals presented to the *Apollo 11* astronauts, on the grounds that they had indeed communicated with Earth from the Moon.

to see and convinced themselves of the canals' reality. Projecting Earth's expectations onto the red planet, they saw evidence of an alien civilization remaking it in our own planet's image, even when no such thing was happening.

By the time the first uncrewed spacecraft were exploring the planets in the 1960s and 1970s, there were few who expected there to be intelligent aliens waiting to greet our robot emissaries. We would have to look farther into the cosmos to find company.

On April 29, 2019, the giant Parkes radio dish in Australia, most famous for being the crucial link that picked up the signal as *Apollo 11* landed on the Moon (an adventure recorded in the film *The Dish*), was pointed toward Proxima Centauri, our nearest star. Proxima is an unexceptional red dwarf, weighing in with about a tenth of the mass of our Sun, and would be of no note whatsoever if it wasn't just next door. Like many nearby stars, it is known to host its own solar system; we have firm evidence for one planet, and there are hints of at least two more buried in the data, but the worlds we know about are unlikely to be very Earth-like. Nothing unexpected should have been found in routine radio observations of this heavily studied star, but analysis of the data coming from the telescope's systems on this particular day revealed a strange new signal. Could it, possibly, have been something artificial?

Radio waves that seemed to be coming from the direction of the star were detected at a very narrow range of wavelengths. The fact that the signal came only in this tight band is important, as most natural sources emit broadly, covering a wide range of wavelengths, rather than copying the kind of narrow-band transmission that's used, for example, to broadcast an old-school radio signal to a set listening at a particular frequency.* This one feature alone made the signal odd, though it seemed real enough, passing simple tests designed to distinguish signals really coming from the sky from ghosts created by the telescope's complex electronics. It disappeared, for example, when the telescope was moved to a different position, and reappeared when pointed back at Proxima. (Such "nodding" is routinely carried out for radio observations, to help pick real signals out of what is often a noisy background.)

The signal persisted for several hours, so it didn't come from a nearby airplane or a satellite traveling overhead. While it was there, it didn't stay at the same frequency, drifting slowly as time passed. This is exactly what we'd expect from a transmitter sitting on a planet in orbit around a distant star; the motion of the planet would cause a change in frequency from a radio transmitter on its surface, just as the pitch of an ambulance siren changes as it dashes past you. The fact that this signal, dubbed BLC1, displayed such behavior was a sign to the astronomers analyzing it, a group from a project called Breakthrough Listen dedicated to looking for intelligence among the stars, that they should take it seriously as a potential alien transmission coming from just next door.

Organized efforts to find radio signals from civilizations among the stars date back to the 1960s and the pioneering years of radio astronomy. The first to propose deliberately trying to

---

*   Do people still remember station frequencies? My mental map has 909 or 693 MHz for sports news in the UK, for example, or 90.7 MHz if you're in New Orleans and want to listen to jazz on WWOZ.

listen for aliens were Giuseppe Cocconi and Philip Morrison, who wrote a seminal paper in 1959 on what became known as the Search for Extraterrestrial Intelligence, or SETI. That paper contains one absolutely fundamental insight that identifies the central problem with such efforts: "The probability of success is difficult to estimate," they note, but "if we never search the chance of success is zero."

Over at Green Bank in West Virginia, a young astronomer named Frank Drake was inspired. Green Bank was, and is, an extraordinary place, set in foothills among the Allegheny Mountain Range. It is remote, and its suitability for radio astronomy (and the military monitoring of signals from satellites around the globe at a separate facility just down the road) is preserved by a "Quiet Zone," a square on the map with sides more than one hundred miles long within which the use of anything that might transmit radio waves is restricted. Close to the telescopes themselves, more stringent rules apply. Wi-Fi is forbidden to the few residents who live in the area, and you won't find a mobile signal no matter how hard you try. No detail is overlooked. A nearby supermarket, for example, has had to use very particular construction materials, preventing signals from its in-house stock-scanning system from reaching the outside world and confounding astronomers. The observatory has patrol trucks to sniff out errant transmissions, though they are slightly more relaxed than they used to be; when I visited, I was told about a recalcitrant neighbor whose illicit Wi-Fi used network names that cast all sorts of colorful aspersions on the region's astronomers and their interference with free, Wi-Fi-enabled living.

ACCIDENTAL ASTRONOMY

When astronomers first moved to this remote place, they took it over from the farmers who had previously worked the land. Drake and his colleagues had their laboratories in a converted farmhouse, known to everyone involved as the Nut Bin. According to the official history of the site, the name comes from the tree that grew behind the house, dropping its nuts into the gutters.* In this unprepossessing accommodation some of the most powerful and innovative radio telescopes ever built were designed and constructed, and for four months in 1960, for six hours a day, Drake pointed one of the largest antennae on the site to look for alien life around two nearby stars.

The observations were known as Project Ozma, named for the queen of the land of Oz, which is, according to the original text of L. Frank Baum's novel, "very far away . . . and populated by strange and exotic beings." I'm not sure Drake or anyone else expected to hear from the Tin Man or any lions, cowardly or otherwise, but the purpose was really to show that scientific attempts to listen for signals from space could be conducted. To improve the odds, Drake chose as his targets Epsilon Eridani and Tau Ceti, both nearby stars that are similar to our Sun. They were, he thought, a good bet as possible hosts of planets. (We learned much later that both do indeed have planetary systems around them; there are at least four and perhaps many more planets around Tau Ceti, while Epsilon Eridani has at least one planet, still embedded in a disc of debris from which more worlds may yet be forming.)

---

* I'm sure this is true. No one would ever consider a bunch of academic astronomers fiddling with wires and roaming the countryside eccentric, after all.

By choosing a radio telescope with which to make his observations, Drake wasn't just making use of the tools that he happened to have available. He also thought that he would be increasing his chances by looking for signals transmitted via a form of radiation that would allow a civilization to send relatively cheap long-distance messages across the galaxy. A glance up at the Milky Way as seen from a clear night sky will reveal dark bands that cut across the glow of light that comes from the stars in the disc of our galaxy. These voids, which have names such as the Cygnus Rift or, in the Southern Hemisphere, the Emu or Coalsack, are regions where nearby clouds of dust—small particles composed mostly of silicon or carbon—block the transmission of optical light. Radio waves are relatively unaffected by such obstructions, so savvy aliens would surely use them to communicate long distance. What's more, in the 1960s the story of technological progress on Earth could be told by explaining that our civilization was becoming radio-loud, sending signals out with an increasing clamor, starting with the first high-powered television broadcasts in the 1930s. Further experiments in radio broadcasts would, it seemed, surely increase the volume.

Following suggestions from Cocconi and Morrison, the Green Bank telescopes were tuned to detect radiation with a wavelength of twenty-one centimeters, the frequency at which neutral hydrogen gas emits radio waves. Hydrogen is the most abundant element in the Universe, and so if you tune a telescope to this wavelength and map the sky, you find that our galaxy shines brightly. Luckily, at this wavelength water in the Earth's atmosphere does not interfere too much with radiation coming from space, and so it was observations made using the twenty-one-centimeter line that first revealed the structure of our Milky Way, with its dense spiral arms standing out clearly even in the earliest maps.

Any alien astronomer with aspirations to understand the galaxy would surely know about the twenty-one-centimeter line and have spent a substantial amount of time using it to map their surroundings, just as we here on Earth had learned to do in the 1950s and 1960s. Drake and company were thinking about finding deliberate communications, civilizations sending out beacons in the hope of being no longer alone. They reasoned that alien astronomers who also had time during a busy research career to transmit a signal out into the cosmos, seeking to chat to their interstellar peers, would be doing so at twenty-one centimeters, surely the safest way of making sure that their counterparts elsewhere, also embarked on the task of mapping the galaxy, would be looking.

Drake was careful to make sure Ozma used existing resources, worried even at this early stage about the possible backlash if substantial money was spent looking for little green men. He was right to be concerned; federally funded agencies in the US, particularly NASA, were forbidden by Congress to spend so much as a cent looking for life for most of the 1990s and early 2000s.

It seemed important to show that observations of this kind were both useful and could be more than an intellectual curiosity. Listening in to observations via loudspeaker* and poring

---

* A common complaint about astronomy in films is that no radio astronomer actually listens to their data come in, though more than a few have been inspired by the iconic scene from *Contact* that shows Jodie Foster's SETI

over the yards of charts that came from the pen of the telescope's recorder, in four months of observations he and the team found very nearly nothing. There was one false alarm, when a plane passing overhead was picked up by the watching astronomers, but nothing of serious interest.

In retrospect, the plane was a godsend. Finding it had at least demonstrated that the search could detect something, if it were out there and transmitting. Ozma inspired a series of increasingly ambitious SETI efforts that have been carried out over subsequent decades. Many have been hindered by a lack of funding for dedicated telescope time, and so what are called commensal programs are increasingly used, where data taken from all over the sky for quite different purposes can be searched simultaneously for anything that might indicate the presence of a signaling intelligence. Before BLC1, the most promising sign arrived in August 1977, detected by the wonderfully named Big Ear radio telescope run by Ohio State University, then operating as a dedicated SETI instrument, following an earlier career spent mapping the Milky Way.

The signal was detected on August 15 and spotted a few days later by Jerry Ehman, a volunteer astronomer charged with reviewing the results of what was then the most ambitious and sustained SETI effort yet attempted. After four years nothing had been found, but Ehman was still there, carefully checking long screeds of printed numbers, each representing the output of the telescope's detector, when he ran across the strongest signal

---

astronomer sitting with headphones on top of her car next to her telescope in the New Mexico desert when aliens finally call. In the case of Ozma, it seems they really did listen.

the project had seen. It was so strong that it seemed it had to be real. It may or may not have been extraterrestrial, but the odds of it being caused by randomly fluctuating noise were practically zero. Ehman circled the result and wrote "Wow!" next to it, thereby giving what became known as the "Wow! signal" its soubriquet, before consulting his colleagues.

They were, and to some extent we all still are, baffled. The Wow! signal was not only strong but also narrow in bandwidth, just as a deliberate signal should be. It didn't seem to correspond to any known satellite, didn't look like any known source of interference, and, though it was still visible when the telescope moved on to the next part of the sky, seventy-two seconds after its discovery it had gone. The signal has never been seen again.

So what was Wow!? Though it remains the most likely candidate for an alien signal, we would be wise to heed Ehman's own words. In a paper written to mark the twentieth anniversary

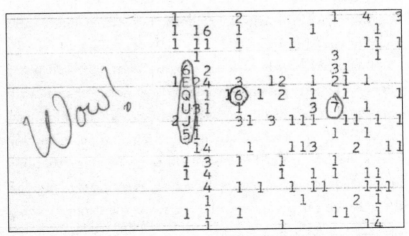

The printout showing data from the Ohio Big Ear telescope; numbers (and the occasional letter) record the strength of signals at different wavelengths and different times. The Wow! signal stands out, with notation by Jerry Ehman.

of the signal's arrival on Earth, he hints at the danger of drawing "vast conclusions from half-vast data," and it's certainly true that without ongoing or repeated observations it is very hard to be sure what caused it. My own brush with explaining Wow! came in 2017 when Antonio Paris, an astronomer at a college in Florida, published a paper arguing that it had been caused by a comet that happened to be passing through the radio telescope's field of view at the time the survey was taken. This got quite a bit of publicity, and I was soon fielding calls from journalists wanting to know whether Paris's claims were right. They weren't—his claimed detection of a twenty-one-centimeter emission from hydrogen in a comet turned out to be an accidental observation of the Sun, which shines brightly at all radio wavelengths—but as I wandered around Oxford that day I tried to work out why the mystery signal couldn't have been a comet.

I started off thinking about comets themselves. These icy snowballs spend most of their time quietly in the frozen outer Solar System, and it is only when their long elliptical orbits bring them swinging closer to the Sun that they become active. The ice sublimates,* forming a coma, or atmosphere, often growing a long tail. A bright comet is one of the most spectacular and beautiful sights in the night sky, made more so by its unpredictability and generally mercurial nature. That's much less true in the radio part of the spectrum, where emissions from comets are feeble, particularly at the twenty-one-centimeter line that the Ohio survey, following the logic of the time, used for their SETI search.

---

* Turns directly from solid to gas, without melting into a liquid state on the way.

Comets bright enough to wow the casual observer are rare, but there are plenty of fainter visitors to our inner part of the Solar System. If every comet were sufficiently luminous to cause something like the Wow! signal every time it got near the Sun, any attempts to do SETI in the radio sky would be filled with false alarms. Instead, comets slip unnoticed among the bright, more distant sources that shine most brilliantly at radio wavelengths. So the Wow! signal could not have been a normal comet, but could it have been a misbehaving one, doing something very unusual indeed?

Maybe. Given that the Wow! signal is something not seen before or since, you can decide it was whatever you like: a radio-bright comet, a distant explosion, or aliens are all, technically, equally convincing explanations. We can, though, compare what a comet capable of causing the Wow! signal would be doing with the behavior of its more everyday siblings. It would have been an enormous outlier, shining tens of thousands of times more brightly than any comet ever observed, so which is the most convincing interpretation: that the Wow! signal was a really, really unusual comet, or that it was something new entirely? I'd say the latter is a more sensible solution, requiring us to believe fewer new things, and so we can put the comet theory to bed.

The Wow! example points to the strategy that SETI has followed for many years: we look for signals that are unusual in some way, such as being especially fast, or confined to a narrow band of frequencies, or repeating rapidly, and if we find them we'll have either a new kind of astronomical object or, well, aliens.

The Wow! signal was a one-off, but what about BLC1? It passed the initial checks, and unlike Wow! was seen on more than one occasion during those first observations in 2019. With some excitement, the team pointed the Parkes dish back at Proxima in November 2020, but found absolutely nothing, a result repeated during the next survey of the star in April 2021. That still left the original detection unexplained, and the Breakthrough team started to think through the possibilities. Could they be picking up transmissions from a satellite in orbit around the Earth? No, they move too quickly and so any signal would drift in frequency more than BLC1 did. What about more spacecraft out in deep space? From our telescopes placed carefully in stable orbits a million kilometers away from Earth, to Juno at Jupiter, and New Horizons and the Voyager probes on their way out of the Solar System, there are a handful of such objects out there, each of them chattering away to controllers back at home. None of the known spacecraft was a good match for the behavior of BLC1, so increasingly off-the-wall ideas were thrown around. What about a transmitter hidden on an asteroid, or an asteroid itself that was somehow reflecting signals from Earth back to us? No dice: the pattern of signals didn't fit an asteroid's orbit around the Sun. Someone on Earth with a transmitter that, accidentally or otherwise, mimicked what would be expected from a celestial source? Not as far as anyone could tell, and how would such a device follow the movement of Proxima across the sky anyway?

At this point, BLC1 had passed more tests than any signal detected by SETI in its long history, and the scheduling of follow-up observations attracted attention. To the chagrin of the team, news leaked to the *Guardian* of what was happening. The resulting report, which caused a great deal of hand-wringing in the Breakthrough team,* is hardly sensational. It includes quotes both from those directly involved in the research and from other independent experts explaining that the signals are "likely interference" and "staggeringly unlikely" to be an alien signal—this latter judgment not least because Proxima's known planet is locked so it keeps one face turned permanently toward its star, as the Moon does to the Earth, making for presumably inhospitable conditions for life.**

The leaking of the discovery of BLC1, and the inadvertent publicity that followed efforts to find out whether it could be a sign of intelligent life, hopefully demonstrates publicly what those of us in the field have long known: astronomers do not keep secrets. The idea of a clandestine network squirreling away evidence of signals in the sky is hard to reconcile with the fact that the most interesting signal found by SETI in decades ended up in the press almost immediately. There are something like fifteen thousand professional astronomers in the world, including PhD students, making us essentially a small village, and news travels fast, especially when telescopes are pressed into service globally to follow some new occurrence in the sky. In 2017, there was much excitement when a global network of astronomers and physicists detected gravitational waves—ripples in space—coming from a

---

* Fair disclosure: I'm on their advisory board and have attended one of their conferences. The dinner was very nice. They have yet to give me millions, or indeed anything, for my own research, though.

** Or at least, for Earth-like life. But there we go again, assuming that our way of living is the only one.

distant explosion that was also seen by orbiting telescopes sensitive to high-energy gamma rays. It was the first time light and gravitational waves were seen from the same event, and pretty much every telescope on and off the planet was quickly swinging to follow it up. Though the detection was secret, the level of activity meant that absolutely everybody knew exactly what was going on. The need to coordinate observations among facilities around the world makes the sharing of an exciting signal inevitable; if there were aliens signaling to us we would all know.

BLC1 did cause a brief press kerfuffle following that *Guardian* report, but in the meantime the team had combed through their data from Parkes in search of similar signals. Examining all of their observations of Proxima from April and May 2019, they found four new detections, though each was fainter than the one they'd spotted first. The weakness of these detections is why they hadn't been found before; hidden in the noisy radio sky, they didn't trigger the software used to hunt for possible signals and were only spotted once the team went looking for signals of the same type as BLC1.

Finding that the signal repeated should have been an exciting moment, further proof that BLC1 was real. Unfortunately, the team also found the same signal fifteen times in data taken when the telescope was pointing away from Proxima, regardless of which of the dish's instruments were being used. On one occasion in particular, the signal persisted as the telescope moved between its target and a position just off-source. That, I'm afraid, definitely makes it interference from something on the Earth: nothing in the sky could follow the telescope as it moved.

The team then looked for similar signals at different frequencies and found that many more examples popped up. BLC1 isn't a unique source but is instead part of a population of annoying chirps, caused by human-made sources emitting at many frequencies at once, interfering with the quest for aliens. To find one unusual signal is exciting. To find many similar signals tells you that you've found nothing more than a new and briefly exciting source of interference.

With BLC1 ruled out, the Wow! signal remains unchallenged as our most likely extraterrestrial source, and then—probably—only because in the 1970s we lacked the wherewithal to record and store enough data to make the kind of careful checks deployed for more modern signals. The work of SETI continues to be arduous and mostly unfulfilled, something that would have been no surprise to Frank Drake and the other pioneers.

The difficulties had been apparent back at the beginning of the search, when Drake gathered a bunch of early enthusiasts at Green Bank in 1961 for an unofficial meeting to discuss the prospects of success for a scientific SETI program.

Things got off to a sensible start. Today, the conference is often remembered for the development of what's known as the Drake Equation, a way of quantifying the number of possible civilizations to talk to. You take your best guess at the number of stars in the galaxy, work out the fraction of those that have planets, the fraction of those that are "Earth-like" in the sense of being able to support our kind of life, the fraction of those that do indeed host life, the fraction of those that host intelligent life,* and the fraction of those that have civilizations, and then multiply the result by the lifetime of such a civilization. The result is a useful estimate of the number of civilizations in the Milky Way that we might communicate with, though it won't produce their phone numbers or instructions for faster-than-light chat.

I interviewed Drake once, and he insisted on calling it "the equation" rather than naming it, as everyone else does, after himself. I'm still not sure if this was genuine or performative humility, but he did agree that "the equation" was essentially a way of quantifying our ignorance. As many of the numbers that go into it, particularly the odds of life getting started and of it becoming intelligent once it does, are still essentially unknown, any outcome—from a galaxy teeming with friends to one whose emptiness condemns us to eternal and inevitable loneliness—can be made to pop out of the equation with sensible inputs.

However, fiddling with the equation does show clearly where our current ignorance lies. When Drake and company were writing it down for the first time at Green Bank, the first few terms, relating to the number and quality of planets in the galaxy on which life might exist, were almost completely unknown. They knew, at least, that the Solar System existed, but, with little idea whether the Sun's family was typical or one in a million, that didn't help much. Almost every term in the Drake Equation was a mystery.

---

* You can use my rule of thumb: an intelligent planet is one with both professional astronomers and gins and tonic.

Remarkably, things have changed dramatically. We have learned in the last few years that planets are extremely common in the galaxy. Go outside and look up, and we are now sure that almost every star you can see will most likely host a planetary system, many of which may contain worlds where conditions are close enough to our own to raise hopes of our kind of life being there.

Our ignorance about the odds of seeing life in the galaxy has passed from being a problem caused by astronomy and our ignorance about planets to a biological one, requiring a better understanding of the origins of life and the likely routes evolution will take once it establishes itself on a planet. Are the living worlds of the Milky Way populated by thinking, social beings, capable of looking up at the cosmos and wondering about us as well as mixing themselves a cocktail at the end of the working week, or are most inhabited planets home to nothing more complex than bacteria? We just don't know.

We're not doing too much better with the rest of the equation. If, one day, those hard questions about life are solved, then as we move from left to right along the equation Drake directs us to turn from biological questions to sociological or eventually even psychological ones: How long will an intelligent civilization last, and why aren't the aliens who live there talking to us anyway?

More than sixty years have passed since Drake started his first experiments, and funding for SETI has ebbed and flowed. For the last few years, private funding from Yuri Milner, an Israeli American hedge fund billionaire who got his start running early internet companies in Russia, has funded Breakthrough's use of the largest and most powerful telescopes on the planet, including those at Green Bank and Parkes, which led to discovery of BLC1. In contrast to earlier searches, which either swept the sky randomly or targeted the odd nearby star, Breakthrough has partly focused on directing the search to those stars that we now know have planets, and in particular those where there is a rocky planet like Earth in an orbit that might give it relatively temperate conditions. That's why they were pointing at Proxima when BLC1 was found.

Despite the decades of effort, so far there is silence, a quiet that has led some to conclude that the odds are hopeless and that we must live in a relatively empty galaxy. The saddest version of this story points out that we have only been a radio-loud civilization for ninety years or so, and that given our track record so far only an incorrigible optimist would believe that humanity will still be here in thousands of years' time. Martin Rees, the cosmologist and Astronomer Royal, a careful thinker absolutely not given to hyperbole, has spent much time considering the risks to humanity in our cosmic and Earthly existence, and gives us only a fifty-fifty chance of making it to the twenty-second century. Even if we survive ten thousand years, that would be a mere moment of cosmic time. The Universe is nearly fourteen billion years old, and so if such a short lifespan is typical, the galaxy will light up with a spark of intelligent life here and there every now and again, only for it to flicker out before there is anyone to talk to. There will be no whoosh of visiting starships or intelligent chatter over the intergalactic airwaves, just the silence of the cosmic dark until we too vanish from the Milky Way, when our time to look out at the Universe and ponder its mysteries is soon over.

Jill Tarter has made SETI her life's work, starting with the marvelously named Search for Extraterrestrial Radio Emissions from Nearby Developed Intelligent Populations—SERENDIP—program that ran at the equally euphoniously named Hat Creek Observatory, continuing with a small band of dedicated researchers through the long, dark period of the 1990s and early 2000s. She is one of the most impressive people I know, with a force of will that reaches much further than one might expect, given her often quiet demeanor. When she speaks, in other words, the rest of the astronomical world listens carefully.

Working in the field has given her a certain monomania, though she would call it focus. I've been in various meetings where discussion of SETI has turned to excited chat about new instruments for radio telescopes, novel machine-learning techniques used to find interesting signals, or the cataloging of nearby stars that might be targets for a search like Breakthrough's. Such things are all useful in astrophysics in general, as well as for SETI, and are a good way of making progress in our search while getting other work done. As the discussion takes off, it is not uncommon to be brought back to what we are supposed to be doing by Jill asking, "Yes, but have we found any aliens yet?"

What's really surprising to me is that Jill has never lost her enthusiasm. She has always known that the odds of success in SETI are unknown, and probably very long, and has persisted anyway. She talked to a group of mostly young astronomers at a conference I helped run in Oxford in 2011, saying that all humanity had accomplished in searching for extraterrestrial intelligence was the equivalent of attempting to investigate the possibilities of life in the sea by examining a bathtub full of water and finding no fish. How absurd it would be to walk away, to choose not to look at the whole ocean in front of us and in doing so continue to explore! What many of us took from that meeting was the idea that we should think, too, about new ways of searching.

Perhaps that was in our minds because of a talk given at the previous iteration of the same conference by Aleks Scholz of St. Andrews University. Aleks spends much of his time considering the behavior of stars and caring for the largest telescope on the UK mainland,* and is one of the most creative people I know. Along with his colleague Kathrin Passig, he has dreamed into existence the WETI Institute, which exists online somewhere in a space among being a scientific research project, a philosophical provocation, and an elaborate joke.

Whereas SETI advocates Searching for Extraterrestrial Intelligence, WETI, working on the basis that civilization on Earth is so recent that any likely communicating races out there are surely well ahead of us, advocates Waiting for Extraterrestrial Intelligence. After all, if those we are hoping to hear from really are much more advanced than us, surely they'll get in touch in their own good time regardless of what we're up to. One option, the WETI Institute website advises, is to place a piece of paper in a book and check it occasionally to see if the aliens have written. For those to whom this seems impossibly low-tech, they also have an Android app that purports to wait

---

* Fun fact: this is the telescope that inspired KT Tunstall's seminal *Eye to the Telescope* album.

several million times a second, saving participants the trouble of doing it themselves.

Of course, WETI's advocates point out that we can't really know how hyperintelligent aliens, whose civilizations have millennia's worth of an evolutionary head start on us, will choose to communicate, and suggest keeping an eye out everywhere. A notable collaboration was with a German T-shirt store that offered visitors to their website the option of purchasing a randomly chosen design; WETI monitored the output of the company's T-shirt generator to see if aliens were trying to communicate with us by altering the apparel of the cool kids in Berlin. The results, Aleks told me, were inconclusive, but regardless of this failure I still think that WETI deserves an Ig Nobel Prize. These gongs, rewarding research that "makes you smile, and then makes you think," are awarded annually in a ceremony that, compared with that for the Nobels in Stockholm, is much less prestigious and certainly has many fewer royals in attendance, but which is undoubtedly much more fun. WETI does make me laugh, but then it makes me realize that we should perhaps look beyond the possibilities of radio signals and think about other ways in which we might stumble upon intelligence in the cosmos.

It's not a novel idea. Paul Davies in his book *The Goldilocks Enigma*, which confronts the awkward question of why our Universe happens to have the properties required for life like ours to exist, calls on us to at least consider the possibility of aliens whenever we find something unexpected in the Universe. It's a thought that underpins many of the stories that I tell in this book, in the hope that each anomaly and odd discovery might be the sign of life that we're waiting for.

Now, though, nearly twenty years after I first read *The Goldilocks Enigma*, I wonder if we can't do better than Davies's suggestion. Rather than wait to be surprised, we can try to imagine what aliens might be doing and go looking for them. One of the most exciting developments has been the advent of new forms of SETI that don't use radio telescopes, looking not for pulses of twenty-one-centimeter radiation sent from distant stars but for flashes of light, either sent deliberately in our direction as messages or used to communicate between alien planets.

As our electronic civilization has evolved, even within the course of a century we've gone from sending messages via broadcast radio to having them travel as light speeding along fiber-optic cables. Maybe light will be our solution once we travel into space: it is easier to transmit large amounts of information this way, as you can modulate the signal faster than you can clumsily change the characteristics of a radio transmission, with its much longer wavelengths.

Luckily, we have become very good at spotting new sources of bright light that shine briefly in the sky. If you visit the observatory at the top of the Canary Island of La Palma, home to Europe's greatest collection of telescopes, almost the first things you'll spot, ahead of the white domes of more conventional instruments that speckle the side of the volcanic island's caldera, are three enormous, shiny mirrors that sit open to the sky. The largest of them, named in the grand astronomical tradition that brought you the Very Large Telescope (VLT), is the Large-Sized Telescope (LST), which is twenty-three meters across, with a secondary mirror hanging forty-five meters above the primary.

The thing is an odd presence under the pitch-black observatory skies, buzzing and whirring as it moves around. It is designed to follow up on powerful explosions detected in the distant cosmos, being able to swing from one side of the sky to the other in just twenty seconds with exquisite accuracy. Instruments like this are designed to spot faint light generated by the arrival in the Earth's atmosphere of high-energy particles called cosmic rays, believed to be produced by supernova explosions, or generated by the powerful magnetic fields that swirl around active black holes. These particle-hunting telescopes normally work in packs, as building an array of them allows for fainter sources to be seen, and for astronomers to quickly rule out false alarms caused by stray light—in La Palma, usually a car headlight as someone travels up to the other observatory domes—so such systems are thus perfect for looking for the flickering of bright, directed lasers.

Again, nothing has yet been found in this new realm of optical SETI, but the LST is a prototype for two large new twin arrays being built, one on La Palma and one farther south, in the Namib Desert, so we should soon be looking for intelligence in the Universe in this way as a matter of routine. A more creative approach still eschews the idea of seeking communication at all, searching instead for the effects that dynamic, spacefaring civilizations might have on their surroundings. My favorite example was work carried out in 2011 by Duncan Forgan, one of the most creative thinkers on SETI, and his colleague Martin Elvis. Casting an envious look out into the Solar System, their collective gaze landed on the resource-rich asteroid belt, home to

millions of rocks filled with valuable rare earth minerals, water
to fuel further travels, or even platinum.*

We have yet to mine our own asteroid belt, but more
advanced civilizations would surely have taken advantage of such
natural resources. Around stars such as Vega, a brilliant presence
in the summer sky seen from the Northern Hemisphere, there
is enough rubble to enable us to study their asteroid belts from
Earth. Duncan and Martin point out that, were we to observe
such a disc with valuable elements missing, it might be a sign
that aliens had been active. It's the astronomical equivalent of
detecting the presence of mice in your house by finding a nib-
bled chocolate bar in the cupboard. As the brightness of a disc
in the infrared depends on the size of dust grains in the system,
they even say that we might be able to detect the side effects of
mining, the star's belt filling with dust as larger rocks are broken
down for industrial uses.

We would need the sensitivity of the telescopes of the future
to attempt this experiment, but it is a good example of how
imagining what aliens might be up to helps us find new ways
of observing the sky. Other astronomers have looked for stars
that glow brightly in the infrared, a sign perhaps that their light
is being absorbed and reemitted as heat by massive orbiting
space stations, and have even examined the properties of whole

---

* There are asteroids so platinum-rich that capturing them and bringing
  them back to Earth would instantly crash the market for the valuable metal,
  rendering your asteroid instantly worthless at the moment of your triumph.
  For exploring the asteroid belt, they should perhaps not send a poet but an
  economist.

galaxies to see if the activities of more advanced civilizations were changing the way they look, perhaps shunting stars around or rearranging things to their liking.

On a smaller scale, there's the possibility of looking for accidental transmissions rather than deliberate signals. If the dreams of astronomers are realized, radio telescopes will soon be sprouting across the wide-open spaces of southern Africa and western Australia. Working together as a single instrument, the Square Kilometre Array (SKA), they should transform radio astronomy, generating as much raw data each year from this instrument alone as passed through the internet in 2017. The result, after processing with a supercomputer, will be the most detailed map of the radio sky ever constructed, and an unparalleled sensitivity to things that change within it.

One thing it might see is emissions from alien radar. Among the most powerful signals we send into space are the transmissions from military radar, booming out in search of incoming aircraft. It has been claimed that SKA, if built over the next few decades to the full extent planned, will be sensitive to airport radar working on any planet within a few hundred light years. Radar, which must cast a wide net to be useful, is one technology where broadcast, rather than targeted emission of lasers, makes sense in the long term, and it may be something that even spacefaring civilizations use regularly. After all, you need to know whether Zog has attracted any hostile company while on their way back from that scouting mission to the next star, and radar will always be a good way of doing that.

The multiple dishes and receivers of the SKA, whose name derives from the total collecting area intended for the telescope's final form, are likely sensitive enough to pick up such a signal from any of the nearest exoplanets. Whether we are sufficiently smart to recover it from among the noise of the rest of the cosmos is less certain. Radar blasts aimed at the sky above an alien spaceport would only occasionally happen to be pointed in our direction, and whether we would be able to recognize these pulses as something worth paying attention to remains an open question.

Just as the Breakthrough team was able to find other examples like BLC1 once they knew it existed, it would be easy to pick out the signal if we knew the specifications of Zog's planet's radar and the frequencies and characteristics of the pulses; but unless there's some strange rule of physics or underlying symmetry that dictates that standards for basic airport equipment are the same throughout the cosmos, that seems unlikely. We might at least expect the signal to occur at the same time each day, but we would need to know how long a day is on the signal's host planet, not necessarily something that's easy to determine when most exoplanets are detected by indirect means, and when we are unlikely in the next century or so to be able even to take a picture of any nearby world.

Whether the SKA does produce an update on alien air traffic conditions or not, the fact that we're arguing about how to do it illustrates the seriousness with which the current generation of SETI researchers are going about their task. We may still be on the shores of Jill Tarter's cosmic ocean, but we have the tools and the desire to scoop at least a few more bathtubs' worth of water out of it in search of alien fish.

For sure, there are always going to be those who are impatient. For some, this means that the glorious dream of SETI isn't worth engaging with. For others, the way to find aliens is to reverse tactics. Rather than trying to listen carefully for signs of life, we should perhaps transmit as loudly as possible in the hope of attracting attention from species who are more advanced and better suited, perhaps, to reaching out once they know that we're here. This isn't SETI, or even WETI, but what my erstwhile colleague Rob Simpson named PETI—the Plea for Extraterrestrial Intelligence, shouting out into the cosmic night in the desperate hope that someone, something, somewhere is listening.

Most experiments in this form of what's usually called active SETI have been essentially art projects rather than anything with a serious chance of success. Among the most famous attempts is the Arecibo message, beamed from a giant radio telescope in Puerto Rico toward M13, a globular cluster on the outskirts of the galaxy. This target was chosen because it contains several tens of thousands of stars, but it is not especially close to us; even if beings on a planet there heard our message, which consists of a crude pictogram containing details of life on Earth,* and responded straightaway, we'd have to wait nearly fifty thousand years for the message to get to us. The point, I think, was to have tried at all, to demonstrate that we could, indeed, attempt to communicate.

In the nearly fifty years since the Arecibo message, occasional sporadic attempts at active SETI have followed, along with hand-wringing by those who fear that in a potentially

---

* Not in a Tripadvisor kind of way, pointing out the must-see sights once our visitors arrive. Instead, the message explains how tall humans are and that our DNA is in the form of a double helix, presumably so the aliens can skip Biology 101 when they get here.

hostile cosmos, attracting attention to ourselves may not be wise. My own opinion is that any civilization with the capacity to do us harm, with all of the capacity for interstellar travel and possession of galaxy-changing tools that implies, must surely also have the ability to find out that we're here, either by detecting our accidental radio emission[*] or, perhaps, by seeing from a distance what we're doing to the atmosphere of this third rock from the Sun.

If they're out there, I suspect they know that we are here. That's probably true even if our neighbors didn't manage to catch the most powerful message ever sent into space, which made use of the EISCAT radar in northern Sweden. The radar is one of the leading facilities in studying the Earth's upper atmosphere, but when it faced a funding crisis a few years ago its scientists realized that they could sell the right to yell into the cosmos to the highest bidder. Quite what any recipients will make of the result, a video advertisement for Doritos, is not clear, but it's unlikely to be a threat to our continued existence, not least because no instructions were included on how to decode the video. Unless they manage to guess exactly how modern web standards work, aliens will be unaffected by the tremendous power of advertising and spared the arrival of a quest across the galaxy with an insatiable appetite for tasty, salted, triangular snacks.

There are many reasons why we have as yet failed to detect aliens. It's possible that, as mentioned above, the spark of civilization burns too weakly in the galaxy to allow anyone to survive long enough to have neighbors to talk to. Or perhaps

---

[*]   I do hope they're enjoying the television of the 1950s and 1960s.

intelligence is rare. After all, evolution didn't work deliberately to produce us thinking monkeys, capable, as Terry Pratchett had it, of sticking our fingers in the electrical sockets of the Universe and fiddling with the switch a bit. For all the good it's done us, it is possible that being smart in general confers no real advantage from an evolutionary point of view. That would explain why it took so long for technological beings to appear on Earth, and why bacteria and other microorganisms get on just fine without needing access to an iPhone. Maybe life itself is rare, and we won the cosmic lottery just by existing on this planet. Or perhaps electromagnetic communication is old hat, and the cool kids of the cosmos are sending each other messages through hyperspace, or by causing stars to blink, or by slinging black holes around to cause gravitational waves. Perhaps, and this is my favorite reason, we are being left alone on some sort of reservation until we grow up a bit and are ready to be part of a grand galactic civilization, an idea called the National Park hypothesis. Or perhaps we have just been unlucky so far, and another year of listening, or the next telescope we build, will pick up the faint signal of our nearest neighbors, themselves reaching into the dark to talk to us.

The possibilities are mind-boggling either way. If life or intelligence is rare, or we are lucky to exist at the brief moment of flourishing that our civilization is afforded, then we look up at a lonely night sky, conscious now that we are the only beings around to witness it, and perhaps the first species ever to do so. We are, in this case, anointed as the part of the Universe that is capable of thinking about itself. On the other hand, a Universe teeming with life, which reduces us to one species in a million, causes my brain to twist in an attempt to place us in our proper perspective. It is very hard to comprehend that we might not be special. What does seem clear is that, in either case, we have a clear responsibility to be alert to surprises, to accidental discoveries. In the first case, where supernovae explode, stars form, or black holes collide, we are the only witnesses to the grandest events in all of creation. In the second, a Universe perhaps richer than anticipated in any sci-fi series* awaits discovery.

---

* At least on television, where budgets mean that alien species tend, where possible, to be bipedal and to be mostly distinguished by having something funny on their heads.

Chapter 2

THE FOUNTAINS OF ENCELADUS

For centuries, those who dreamed of making contact with aliens thought about life not among the stars but on the planets that orbit with the Earth around our friendly Sun. For most of astronomical history, it would have seemed natural that worlds that share their basic characteristics with Earth would be homes to life. Percival Lowell (see p. 29) certainly believed that Venus and Mars in particular were inhabited.

As humanity continued to explore the Solar System, sending out robotic probes from Earth to prod, poke, and image our neighboring worlds, it became obvious that it was increasingly unlikely that our first encounter with an alien species would occur on the desert surface of dusty Mars, for so much of history considered the most likely home for life beyond Earth. Nor would I bet on any of the other rocky worlds that cluster with Earth around the Sun. Instead, if I had to put money down, I'd gamble on news of the discovery of the first alien biota coming from a small, sophisticated amphibious craft, slowly melting its way through the icy crust of Enceladus, a tiny moon in orbit around Saturn, before swimming free in a salty ocean of liquid water. Here, far from the warmth of the inner Solar System, but in an environment heated by tidal forces we do not yet fully understand, a mode of aquatic life may have evolved and flourished, both eerily familiar and yet utterly different from that which teems throughout the seas of Earth.

Enceladus was discovered in the eighteenth century by William Herschel. He saw the moon only as a tiny point of light that kept Saturn company as the planet moved across the sky. In addition to its magnificent ring system, Saturn is accompanied in its orbit by a great retinue of more than a hundred known moons, and most of them, including Enceladus, remained almost entirely obscure until the system was visited by the twin Voyager spacecraft in 1980 and 1981. Images sent back from this first encounter revealed that Enceladus, which is only about the size of the British Isles, has a surface consisting mostly of bright white water ice, a feature that made this moon the most reflective object ever found in the Solar System. The images the probes returned did not cover the whole surface, leaving blank spots in even our best maps; but even from what was reported from these brief flybys it is possible to see that half the moon is smooth, the ice broken by just a few odd, sinuous features that streak across an otherwise monotonous surface, while the other half is clearly different, marked by numerous large craters.

Spotting craters anywhere in the Solar System is both exciting and useful. More than four billion years since the planets formed, there is still a lot of leftover rubble and debris around. The old trope of an asteroid belt dense enough to hide a ship in, a staple of the science fiction I was raised on,* may be unrealistic, as even in the most congested bit of the main asteroid belt that lies between Mars and Jupiter there will be, on average, nearly a million kilometers between neighboring rocks of any size at all. Yet given enough time, collisions between smaller bodies and larger targets, including asteroids hitting the Earth, become inevitable.

On our own planet, most incoming objects, particularly the smaller ones, burn up in the atmosphere and do little damage. If you have ever seen a shooting star, then you have witnessed the destruction of a particle the size of a grain of sand, immolated in the upper atmosphere ninety kilometers or so up, where much of its material will remain. The resulting layer is of great use to astronomers, who fire lasers at it, not out of some misguided attempt at planetary defense but because by doing so we can cause the sodium in the meteor layer to glow, creating artificial stars. These are used by modern telescopes to measure and then adjust for the twinkling of the real stars, caused by the passage of their light through Earth's atmosphere. In this way we can simulate the effect of putting our largest telescopes into space, avoiding the distorting effect of the Earth's atmosphere altogether.

---

*   So. Many. *Star*. *Trek*. Episodes.

Not everything is trapped in the upper atmosphere. An estimated one hundred tons of space dust does make it to Earth's surface each year, settling slowly onto the ground, which means that if you have neglected to clean your gutter recently it likely contains at least one grain of interplanetary material, though it may not be easy to find.

Most things larger than a dust grain will burn up on the way to the surface. Because of the thickness of the Earth's atmosphere, only the largest incoming bodies have a hope of reaching the surface, but they do make it occasionally; one, for example, fell through a roof and neatly plopped into a laundry basket belonging to a resident of Park Forest, just outside of Chicago, in 2003. But to find more impressive craters, we need to go to the deserts of the world, where evidence of previous impacts is less likely to have been weathered away.

I remember standing in the middle of the northern Arizona desert, looking down at the stark puncture wound that is Meteor (sometimes called Barringer) Crater, a hole more than a kilometer across and 170 meters deep that was punched into the ground maybe fifty thousand years ago. That it is the product of a meteorite impact is beyond doubt, as iron from the impactor has been dug up from the crater floor, and much still lies underground. Yet the crater's prominence is as much due to its location in a dry desert environment as to the force of the blow that produced it. Almost anywhere else on the Earth's surface, a crater like this would have been quickly eroded by the joint efforts of wind, rain, rivers, and, just as importantly, life.

Look up at the Moon on any clear night and it is clear that this process of erasing cratered history happens more slowly on the other worlds of the Solar System. Even with the naked eye, it is possible to see that the Moon has craters, especially if you are viewing on one of the nights either side of the full Moon. At this time of the month, the noonday overhead Sun catches bright rays of debris scattered by impacts that formed the largest impact basins. Two in particular stand out. In the middle of the lunar disc is Copernicus, and Plato sits near the Moon's southern limb, both of them surrounded by rays of bright material that stretch across the lunar surface. We know that these rays are the result of impacts because a team of Japanese researchers have made smaller versions by firing projectiles into a lumpy and uneven model of the Moon's surface, spraying material everywhere.* Even a small pair of binoculars will show many more craters at any phase, and a moment spent looking at the Moon with a telescope will confirm that the Earth's satellite is a world whose landscape of flat seas and rugged highland areas has been shaped by impacts.

At the top of the staircase in Oxford's History of Science Museum hangs an extraordinary eighteenth-century drawing of the Moon made by the portrait painter John Russell, pieced together from painstaking observation of the lunar surface at the highest magnifications then available. In Russell's hands the surface of the paper seems to spring into life, with thick layers of pastel almost smeared onto the black background, creating a three-dimensional effect. As your gaze travels over the drawing,

---

* I once visited a team at NASA's Ames Research Center in California who were doing similar things and was most gratified from pictures on the walls to see that they'd clearly tested the effect of firing a high-energy bullet into a landscape populated by plastic dinosaurs.

past the ragged, cratered, and brighter highlands, you encounter the sudden smoothness of the dark patches of "*Maria*," or seas, whose shape forms the "man in the Moon" (or, if you prefer East Asian or Native American legend, a rabbit or hare). These flat plains are more friendly to an explorer than the mountainous terrain found elsewhere, which is why all but one of the *Apollo* missions that landed on the Moon headed for the seas, where craters, though present—Buzz Aldrin had to steer the Eagle landing module away from one just before *Apollo 11*'s historic touchdown—are rarer.

Why is there this difference? Incoming asteroids are not coordinating their assaults on the lunar surface, choosing which parts to aim for before landing in formation only on the highlands. Instead, the difference in the number of craters can be easily explained by assuming that the seas are simply younger than the rest of the surface. Such features are thought to be originally volcanic seas of lava that spread out from vents, pooling in the lower parts of the lunar surface and cooling to produce the flat plains that we see today. As time passes the occasional meteorite will hit, and without a thick atmosphere to cushion the blow, or significant erosion to hide the slowly accumulating traces, the surface of these frozen lava seas will begin to change. The simple rule is that, all else being equal, a surface with more craters is older than one with fewer.

This idea had been around for a few decades when Armstrong, Aldrin, and the rest of the *Apollo* astronauts took their small steps on the lunar surface. The first thing they did upon leaving their spacecraft was to bag up rocks and a first set of soil samples, so that even in the event of problems causing an early abort they would still bring something home for the laboratories of the waiting scientists. Over the course of increasingly extended sojourns on the Moon, in later missions aided by the ramshackle but effective lunar rovers that greatly extended the area that could be explored, astronauts added to these grab-bags with more judicious selections. When they splashed down in the Pacific, they brought with them kilograms of rocks from our sister world. The customs manifest filled out by the *Apollo 11* astronauts, reporting their trip from origin "Moon" to destination Honolulu, includes a cargo reported as "Moon rock and Moon dust samples."* Once in terrestrial laboratories, careful analysis of the composition of the samples allowed for precise dates of formation to be given to the six sites that had been visited. As it turned out, rocks collected by *Apollo 16* from the cratered Descartes Highlands were older than those taken during *Apollo 11*'s visit to the flat Sea of Tranquility.

Even better, by using these results to calibrate how we convert crater counts to geological age it has been possible to date much of the lunar surface. In fact, the crater record on the Moon provides the page numbers for the entire history of the Solar System, telling us what happened in what order, as emboldened planetary scientists have applied the same crater-counting techniques far and wide. They do need to worry about differences in the incoming flux of asteroids in different places, in that we should expect a different rate of impact on, say,

---

* Buzz Aldrin also claimed $33.31 in travel expenses, based on the mileage he'd traveled. Legend has it that he was threatened with being charged for damage to the vehicle, most of which was left by design on the Moon or burned up in the Earth's atmosphere.

a moon of Saturn than on our Moon, hanging out in Earth's orbit, but such things can be modeled. Anchored in ground truth obtained from humanity's six fleeting visits to the lunar surface, the simple act of counting craters should let us come up with an estimate of the age of any surface in the Solar System.

Small worlds like Enceladus should be dead. The Earth is still being heated from within by the decay of substantial amounts of radioactive material swept up as it was forming, which subsequently sank to its core. The heat produced fuels plate tectonics and the dramatic volcanoes scattered around the Earth's surface. Though Venus doesn't seem to have continents that drift like ours do, it almost certainly has active volcanoes that are more than capable of resurfacing the entire planet, covering it in fresh lava every few hundred million years. (As we'd expect, few craters on Venus are known, though the thick sulfurous atmosphere through which any incoming meteoroid would have to travel must play a role here too, burning up all but the most massive rocks.) Most large worlds are similar, but Enceladus is small, and its supply of this ancient heat must have long since been exhausted, so the lack of craters on half of its surface in the Voyager images were an early sign that something unexpected was happening.

An initial flurry of interest in strangely craterless Enceladus prompted a variety of possible explanations, mostly involving odd orbital mechanics that placed the tiny moon on paths that would occasionally send it swinging closer to Saturn, producing tides capable of melting, at least partly, its outer surface. Tides induced by the giant planets may be significant in the lives of many of their satellites; Jupiter's moon, Io, owes its status as the most volcanically active place* in the Solar System to an orbit that continually stretches it in Jupiter's strong gravitational field. If Enceladus swung regularly close to Saturn, it would be squeezed and heated like a squash ball bouncing off the wall, keeping its surface smooth, but it is hard to explain why Enceladus's orbit, today perfectly stable, may have produced such an effect in the recent past. More ideas were bandied about, with the headline "Could Saturn's Rings Have Melted Enceladus?" appearing in the news section of *Science* in 1984;** but the problem was soon forgotten, set aside like many mysteries of the outer Solar System until the next spacecraft could visit.

After a cruise of nearly seven years following its launch on a Titan IV rocket, NASA's Cassini probe arrived at Saturn in 2004. Its first task was to drop the European Huygens lander off at Titan, Saturn's largest moon and the only satellite in the Solar System to have a thick atmosphere. As Huygens floated down on its parachute through the moon's hazy clouds its instruments revealed a world shaped by methane rain, with rivers, gullies, and, we would later discover, hydrocarbon lakes and seas. Titan, with the rich chemical stew of its atmosphere and surface, is a fascinating place, and perhaps one where chemical cycles just short of

---

* Earth, we think, is second, but it does not come even close.
** A rare scientific example of what political columnist John Rentoul calls a QTWTAIN—a Question To Which The Answer Is No, a phenomenon beloved of headline writers forced to wring drama out of uncertainty. It's always "Could Aliens Walk Among Us?" in the tabloids, never "Aliens Walk Among Us!"

the complexity needed for life may exist,* but the attention of the Cassini team was soon diverted to an unexpected target.

The geometry involved in setting up the Titan encounter meant that on February 17, 2005, Cassini flew just 1,170 kilometers above the surface of Enceladus.** Detailed images were taken and returned to Earth, pleasing the watching team with higher-resolution views than had been possible from Voyager, some of which showed pristine ice looking as pure as freshly driven snow and others craters, albeit ones blurred and worn away by an unknown process. They looked very different from those on the Moon, for example, where even billion-year-old craters return to their sharpness and distinctiveness in the absence of any atmosphere.

The images were interesting, but were more valued at the time as proof that Cassini's cameras were working well early in the spacecraft's stay at Saturn than as a sign that Enceladus would be worth a second look later in the mission. The team behind Cassini's magnetometer, an instrument capable of sensing subtle variations in Saturn's magnetic field, were also paying attention, hoping to use the flyby as a test of their new toy's capabilities. The magnetometer is slung out on a long boom that protrudes from the spacecraft, keeping it clear of interference from the electronics on board, its primary duty being to help understand how the solar wind—the flow of particles from the

---

*   I am very excited indeed by the upcoming Dragonfly mission, which early in the next decade will land a helicopter on Titan and use it to explore the diverse landscape of the moon's surface.

**   Essentially a coincidence, though one anticipated and planned for, rather like my frequent, completely spontaneous encounters with the fish and chip shop while cycling back into town from tennis.

Sun that fills the Solar System—affects and interacts with Saturn. It was already known from Hubble Space Telescope images that such particles were often channeled down toward the planet's extremities, exciting atoms in the atmosphere and creating spectacular aurorae that dance around the poles, just as we have the Northern and Southern Lights here on Earth.

Such displays add to Saturn's status as the Solar System's premier tourist attraction for the discerning traveler for whom distance is no object,* but when we try to understand them in detail they also expose gaps in our understanding of how giant planets work. The magnetic field even plays a role in how the planet rotates, and as a result, although Cassini's seventeen-year-long mission is complete, we still have no clear idea of how long a day actually is in Saturn's complicated, windy atmosphere.

None of which is to suggest that anyone expected the magnetometer to do anything interesting as Cassini's passage swept it through space as the spacecraft passed Enceladus. Nonetheless, the instrument team, excited by exploring the worlds they had dreamed of through long years of writing proposals for funding, instrument-building, and then the lengthy snooze during the spacecraft's journey to Saturn, had their instrument on and took the trouble to review and analyze what it had recorded during the flyby. To everyone's complete astonishment, both during this first visit to Enceladus and on a second, slightly closer pass less than a month later, data collected by the magnetometer suggested that Saturn's magnetic field seemed to be

---

* Bored by the rings? Fancy something a little bit different? Swoop over the poles and see Saturn's most spectacular light show . . .

bending away from the planet to drape itself around the tiny moon. Something on, or around, Enceladus must be interacting with the field, and were it not a patently ridiculous suggestion for such a small body, the most usual explanation would be that the little moon had some sort of substantial atmosphere.

That idea made no sense. As noted above, of all the Solar System satellites only Titan, bigger than Mercury and nearly as massive, and benefiting from the cold environment of the outer Solar System, has a substantial atmosphere. Titan is so big that if it orbited the Sun, instead of Saturn, we'd most likely call it a planet,* and its gravitational pull is enough to retain its thick atmosphere; but Enceladus, at just five hundred kilometers across, was surely too small to copy its larger sibling. Its gravity should have been insufficient to cling to any sort of atmosphere. And yet the data seemed to say otherwise.

What was going on? Intrigued, the woman in charge of the magnetometer, the indefatigable Michele Dougherty from Imperial College London, persuaded the rest of her Cassini colleagues that they should fly the spacecraft as close as possible to the surface of Enceladus. Changing plans early in the mission could not have been easy, and it's a tribute to Michele's persistence and the respect for her science that on July 14, 2005, Cassini passed just 175 kilometers above Enceladus, much closer than any spacecraft had ever been. In doing so the probe gave itself the first interplanetary car wash, passing right through what turned out to be jets of water shooting into space from the south pole of the moon. Had anyone involved known what it was flying into, such

---

*   Or at least still be arguing about it, like poor Pluto.

a joyride would surely never have been authorized, but, blessed with ignorance, they sent Cassini skimming through what would become known as the fountains of Enceladus.

This close flyby was only taking place because Michele and her colleagues paid attention to what they fully expected to be routine magnetometer data, but the results obtained during the probe's close passage would promote Enceladus from an after-thought to a major focus of Cassini's mission. It made many subsequent visits to the mysterious moon, though none quite so recklessly close. During these trips the spacecraft's cameras returned spectacular images showing the fountains' origins in multiple individual jets scattered along the moon's "tiger-stripe" features, as the branching and innocuous-seeming linear features originally spotted in the Voyager images were now known.

Enceladus seen in silhouette with the rings of Saturn in the foreground in this spectacular image from Cassini. The fountains, which are responsible for the tenuous outer ring, are visible at the moon's south pole. In the background, the tiny moon Pandora is also seen. *Credit: NASA/JPL-Caltech/Space Science Institute/PIA17144*

Yet the most interesting discovery would once again be made by one of the less celebrated instruments on board.

Astronomers love dust. Whether we're considering frigid star-forming regions or worrying about the complex dynamics of Saturn's rings, understanding the behavior of these tiny grains of silicon or carbon, each perhaps a tenth of the size of a grain of sand, turns out to be rather important. Cassini therefore carried a Cosmic Dust Analyzer, a rather grandiose name for what was essentially a bucket dragged around by the spacecraft, designed to catch and then analyze* such particles throughout the tour of the Saturnian system. The particles this sensitive chemical sniffer found while traveling through the fountains turned out to taste salty, like a proper ocean. Whatever the source of the water that streams from Enceladus, it must somewhere be in contact with a seabed or river floor, and that in turn means that the moon must have the potential to be a true habitat for life. Rather than being an inert, ice-encrusted ball of water, there is, somewhere inside Enceladus, a place rather like the one in which life may have got started on Earth.

The last time we found such a place on a planet was in 1977. A group of ocean scientists and geologists aboard the research ship *Knorr* spotted something remarkable in images being relayed from their underwater research platform, which delighted in the name of ANGUS. Their mission was to image the seabed some 2,500 meters below them, to help us understand how the slow grind of plate tectonics reshapes the Earth's crust. Finding a

---

*   OK, the analysis bit upgrades CDA past your standard beachside bucket, but the principle is the same.

glow of light in this deep ocean, coming from a region of super-heated water, was a complete surprise. Drawing closer, the team found a diverse variety of unusual and spectacular life-forms clustered around a small vent in the ocean floor. The inhabitants of this strange place ranged from enormous clams and worms to a surprising range of plankton. Each was a new species, but so unexpected was this find that the *Knorr*'s complement of thirty or so scientists included not a single biologist.

Future expeditions, which rapidly discovered more of these features and quickly became known as hydrothermal vents, did not make this mistake. These places—where cold seawater seeping through cracks in the ocean floor suddenly encounters rocks heated by rising magma from the mid-ocean ridges where new seafloor is being made—can reach temperatures as high as 400 degrees Celsius, driving all sorts of chemical reactions and liberating minerals from the rocks, turning the normally inert deep ocean into a nutritious soup. While almost every other living being on Earth gets its energy, directly or indirectly, from the Sun via the process of photosynthesis, the beings growing, scuttling, and swimming around these mid-ocean oases draw the energy they need directly from the center of our planet.

In some ways, it is a simpler life. Photosynthesis—the chlorophyll-based trick by which plants and many bacteria convert sunlight into something useful—is undoubtedly complicated. Although the fact that it occurs in organisms as different as a sequoia tree and a tiny bacterium does suggest its origins lie at the root of life's evolutionary tree, the chemical pathways needed to turn sunlight efficiently into energy may just be too complex for life's first attempts. Instead, many have come to believe that the first living organisms found themselves huddled around a warm patch in the middle of the ocean, where high temperatures have triggered the complex chemical reactions that lead to life.

Could such a place exist—and harbor life—under Enceladus's icy shell? Toward the end of its long exploration of the Saturnian system, Cassini returned to Enceladus. In an October 2015 flyby, it made what might be the most interesting discovery yet. Having learned from previous encounters how to really get the most out of the instruments on board the aging spacecraft, its scientists were able to look not only for water and salts, but for the presence of gas trapped within the water. They found both hydrogen and carbon dioxide, each in concentrations that accounted for a percent or two of the whole jet, more *pétillant*, or slightly sparkling, than the full seltzer, but enough to suggest that the seafloor under Enceladus's ocean must be an active place. The most likely source for those gases is a vent just like those in Earth's ocean, powered not by upwelling magma, as on our planet, but by the flexing of the seabed under the varying pull of Saturn as Enceladus moves around its slightly eccentric orbit. The structure of the moon amplifies these subtle changes, meaning that the tides can have a spectacular effect on its interior without sending it careening off its normal orbit.

This doesn't yet mean there must be life. As Drake would remind us, we do not understand in any sort of detail what conditions are needed for life to get started, nor what the odds are of it doing so if such conditions are provided. The potential origins of life on Earth, with our distant, distant ancestors clustered around a hydrothermal vent, are still speculative, although well informed. We also don't know what the odds are of producing something even as complicated as bacteria once life has, somehow, got started. It might also take a while for the right ingredients to assemble, or for two molecules to get lucky and become a self-replicating system. A crucial question is thus, how long have the oceans of Enceladus existed? If they are long-lived, then the chance of life, and even complex life, existing there surely must increase.

Has Cassini been investigating a long-lived phenomenon, with Enceladus's ocean a feature of the Saturnian system for hundreds of millions or even billions of years, or have we happened by at exactly the point where a recent disruption to the moon's orbit, or some other factor, has triggered the observed activity? A similar discussion has been raging[*] for years about Saturn's magnificent rings, which some think date back to the time when the planets were forming and others feel must be the result of some recent cataclysm, like the sudden capture and destruction of a rogue moon, so that they are a temporary, if spectacular, addition to our Solar System.

Despite their appearance in a telescope on Earth, the rings are not solid but consist instead of many millions of icy fragments, each of which orbits within a thin disc no more than twenty meters across—less than the height of a five-story office block. This remarkable structure is maintained partly by the angular momentum of the ring particles themselves, but also through the gravitational attention of a set of so-called shepherd moons. As these bodies, some of them embedded in the ring system and

---

[*]   Politely. Mostly.

others just outside, interact with the surrounding particles, they create gaps in the rings, places where no stable orbit exists, some of which, like the well-known Cassini division, are large enough to have been discovered through telescopic observation several hundred years ago.

In cataloging these divisions and beginning to understand the complexity of the dance performed by the ring particles as they orbit Saturn, astronomers noticed an outermost ring, different from the others, fainter and gossamer-thin, and quite separate from the inner structures. This is the E-ring, and it is odd. It is not as flattened or as clearly defined as the other rings, more of a softly glowing doughnut than the sharp, crisp rings I can see with my small telescope from the back garden. This ring is, we now know, the result of the activity on nearby Enceladus, pouring water into space, creating countless sparkling ice crystals that spread out around the planet, reflecting sunlight back to observers on Earth. Given its brightness and size, if it is formed completely from water from the moon then the E-ring suggests that Enceladus's ocean might be long-lived; the moon and its fountains must have been there a while to create such a broad feature.

Further progress has also been made in trying to understand the source of the energy that keeps Enceladus's ocean liquid. The gentle squeezing of the moon on its thirty-three-hour orbit around Saturn is, once its interior structure is taken into account, enough to keep a subsurface ocean going indefinitely, and its orbit seems stable enough for its current situation to have been unchanged for at least a billion years. The tiny moon really does seem to be a long-lived potential home for life.

Enceladus is special. It is the only place in the Solar System where we have direct access to material from a potential habitat without having to go to all the bother of landing, drilling, or digging to find it. If life still exists underneath the barren and bleached deserts of Mars, then we are going to struggle to locate it. A floating chemical laboratory, far beyond our capabilities to fit onto a spacecraft, would be needed to properly understand the hydrocarbon lakes of Titan, and any mission digging into Martian soil looking for life will have a very tough time of it. A decent guess at whether life is widespread in Enceladus's oceans could be made by sending a craft equipped with a suitable suite of instruments back on Cassini's path, passing through the plumes and filling a sample bucket with fresh Enceladian water as it does so.* The presence of the fountains, as well as the evidence of recent resurfacing, also reassures us that the ocean cannot be far beneath the ice shell. The water is right there, where we could get to it. Ambitious missions might someday land on the moon's surface, drill through the ice, and explore the oceans directly, but there is no need. We can just sample what Enceladus is already spraying into space.

Excitingly, it turns out that what renders Enceladus unusual isn't the presence of its under-ice ocean but the fact that the slow release of water from it into space makes it easy for us to access

---

* I have wondered if there's a commercial case for selling this as the most exclusive mineral water in the Solar System, but quite apart from the ethics of encouraging the super-rich to mix ice cubes made from alien water sources with their single malt (it might affect the whisky, after all) the costs of a return trip to Saturn are currently prohibitive.

it. Other worlds hide their liquid water from us. At least two of Jupiter's large moons, Europa and Ganymede, seem to have substantial subsurface oceans, their presence inferred from studies of the moons' craterless surfaces, which betray recent activity, and which contain icy forms that look for all the world like floating icebergs caught in a winter freeze somewhere in the Antarctic. The Galileo probe, which did for Jupiter and its multiplicity of moons what Cassini did for the Saturn system, provided spectacular images of both, and the smart money in both cases is on the premise that the oceans are deep—perhaps one hundred kilometers—under the surface. Observations with the Hubble Space Telescope do seem to suggest that Europa, at least, has occasional episodes when water plumes are visible above weak points in the ice, but they are not as persistent or spectacular as those seen around Enceladus and it's not clear whether they are coming from a water source connected to the main ocean or some separate surface phenomenon.

More recently, Europa has been visited by the Juno probe, which has been in orbit around Jupiter since 2016, swinging low over the planet's poles once on every orbit to map its storms close up. Each trip has produced beautiful and fascinating images of the swirling storms in its atmosphere, seen closer than ever before. In one recent image, of Jupiter's southern pole, we look down on the dark pole of the planet surrounded by no fewer than seven major cyclones and an atmosphere in which turbulent forms whirl and dance endlessly.

Juno nearly carried out its assignment without the ability to take any images at all. Once NASA had selected the mission, a probe with the goal of investigating Jupiter's environment, the possibilities were legion. Some scientists wanted to study the giant planet's magnetic fields, while others dreamed of understanding heat flow in the atmosphere. The powerful radiation belts, the equivalent of Earth's Van Allen belts, needed monitoring to understand how Jupiter interacts with the solar wind, and an ultraviolet sensor would be essential to keep an eye on the planet's spectacular aurorae. Tracking its movements by using the probe as a radio beacon would allow Jupiter's gravity field to be mapped and so its interior to be understood. For the first time, we'd know if there was a solid core deep down somewhere under the thick atmosphere and the clouds that pass for a visible surface, perhaps the remnant of an Earth-sized world that had formed previously.

Over time, a suite of instruments was selected for the purpose, taking careful account of the amount of power, time, and data each would consume, and—of course—how many dollars would be needed to build them. When the job was done, a general-purpose optical or infrared camera was missing, considered to be good for sightseeing but less than vital for the scientific mission the Juno team had in mind. Which is all very sensible, but surely if you're going to go somewhere exciting—particularly somewhere never visited before—you pack a camera.* Common sense thankfully prevailed, and Juno's instrument suite gained the snappily named JunoCam, placed on board with the specific goal of increasing public interest by sending postcards from its journey. Any glance at the images it has returned will show

---

* Or, yes, a mobile phone, but at Jupiter the roaming fees are likely extortionate.

you it has done much more than that, providing vital context for the other instruments and, with the sheer complexity of the dramatic storms and swirling clouds it revealed, raising a wealth of questions of its own.

In September 2022, Juno's looping path took it close to Europa, passing just 360 kilometers from the moon's surface, providing spectacular new images. The highest-resolution image JunoCam sent back showed an icy surface covered in grooves and ridges, dimly lit by the reflection of light from Jupiter, which illuminated the nighttime surface as the spacecraft rushed by. The structures in the image seem to be telling us about a complex history of activity that created them, and some dark splotches[*] appear to represent places where material has been sprayed up and onto the surface.

To make sense of what we're seeing, we need to take a look at the whole surface. Two new spacecraft, NASA's Europa Clipper and the European Space Agency's (ESA) JUICE, are to follow in Juno's path, exploring these icy worlds in the next decade with the primary aim of mapping the extent and depth of their oceans. JUICE, which will visit Europa and the outermost large moon, Callisto, before conducting a longer study while in orbit around Ganymede, is already on its way. Elsewhere in the Solar System, the pattern seen on Enceladus and Jupiter's moons, with their internal oceans, seems to recur—although with more limited evidence—on several moons of Uranus, on Triton, Neptune's largest satellite, and perhaps even on icy Pluto, which appears in images from the New Horizons spacecraft to have some smooth,

---

[*]   Not everything needs a technical term.

craterless regions too,* though they may be refreshed over millions of years via the action of the slow, cold equivalent of a lava lamp.

Ocean environments beneath an icy surface may even be the most common homes for life in the galaxy. Living on the surface of a planet like the Earth requires a temperate climate, at least if creatures like you and I are to survive, and that means selecting worlds that happen to be just the right distance from their stars. Of all the worlds we know of in our Solar System only the Earth-Moon pair are within this habitable zone.** If you're prepared to live in an ice-capped ocean, you can exist wherever you like in a Solar System, warmed by the renewable energy of the tidal forces provided by your moon's parent planet's gravity rather than being reliant on sun- or starlight.

As we learn about other planetary systems, it seems likely that giant planets throughout the Milky Way might be accompanied by large, icy moons. Based on what we see in the Solar System, many of these moons appear to have internal oceans. If these are indeed habitats for complex creatures, then their denizens may consider their way of life, enclosed in an icy shell, as the normal state of existence and consider those of us who live out in the open, on the surface of a world and able to look up to a sky filled with stars, as freaks. Our kind of life may be in a minority in the Universe.

---

* The largest of them, now named Sputnik Planitia, was adorably heart-shaped. Excellent marketing, Pluto.
** Or, as it used to be called before exoplanet researchers got serious, the Goldilocks Zone, where the temperature is neither too hot nor too cold, but just right.

Life for such aquatic beings must be very different from our own. Earth's dolphins suggest that intelligence may be possible in a marine environment, though dolphins, like all whales, did spend some of their evolutionary past as land-dwelling mammals. No matter how smart they are, without easy access to the night sky Enceladean cetaceans are unlikely to have produced many great astronomers. Their physicists may have noticed the tidal effects of their world's orbit, and if they understand Newtonian gravity (a force of less importance in a buoyant world than it is in our own landlubbers' existence), they may have been able to deduce that they live in a world that orbits another, more massive, planet. Perhaps we can imagine them one day setting out to see it, drilling cautiously up through the icy shell that surrounds their home ocean, just as we have considered visiting their home by drilling down.

Emerging into empty space for the first time would be a shock surely greater than that of seeing the Earth from orbit, creating a jet of water sparkling as it was first exposed to sunlight.[*] We can imagine the alien astronauts beating a hasty retreat, amid much discussion of what the strange, waterless Universe outside their home might be. Eventually, a craft—a sphere of water, protected somehow against the vacuum of space[**]—might emerge, and its occupants would become the first of their species to see Saturn in its glorious ringed magnificence.

---

[*] I am not suggesting that the fountains of Enceladus are signs of traffic to a busy, icebound, alien spaceport. I'm merely saying that if such an alien race did indeed send ships to and fro through fissures in their world's icy crust, we might expect the odd jet of water to be lost.

[**] A bubble car, if you will.

Or perhaps not. Any species that has evolved over millennia in the darkness of an interior ocean is unlikely to have much use for a sophisticated sense of sight. An alien environment will surely in reality produce alien creatures, with modes of thought and ways of life, and our struggle in understanding Enceladus may be in trying to fit its possibilities into human experience, rather like expecting a Saturnian version of Flipper the dolphin to spin us a somersault of welcome if ever we meet. Beings not reliant on sight will develop other senses, one assumes, and given the story of how we discovered Enceladus's oceans in the first place there's something glorious about imagining its inhabitants being sensitive to magnetic fields.

Speculation and the wilder shores of exozoology aside, Enceladus is clearly a fascinating world, and mysteries remain. It now seems clear that the tidal action of Saturn is the cause of the activity we observe, but it is not apparent why this seems to affect the south more than the north. While the south pole has the fountains and tiger stripes, the north is quite different, with terrain covered in craters that distinguish it from younger material on the other side. One early explanation, which suggested that the ice at the north is thicker than that at the south, has already been ruled out. Though the ice is indeed thicker, it is still thinner than that away from the poles, and jets of water should still be able to punch through. This puzzle, and a desire to understand the size and character of its ocean, taking one more step toward our possible encounter with life, will surely take us back to Enceladus soon. This time, as a life-seeking spacecraft attempts its first flyby of the fountains of Enceladus, absolutely everyone will be paying attention.

Chapter 3

THE SCOUT FROM REALLY, REALLY FAR AWAY

When it finally arrived, it caught the astronomers of Earth off guard. It's true that some of them had expected a visitor like this one, emerging suddenly from the darkness of interstellar space and plummeting toward the inner Solar System, but the equipment at their disposal was too primitive to show the interloper as much more than a point of light. In truth, they hadn't expected to be able to spot something this small, its surface blackened from its long journey between the stars. No one was quite sure how to react as they watched it, and arguments—about what shape the mysterious visitor was, where exactly it had come from, and where it might be going next—broke out around the world as sky watchers everywhere pored over what little data they could glean from their telescopes.

Even if we humans don't understand something, we still name it. This first interstellar visitor was named 'Oumuamua, a word constructed from the Hawai'ian language to honor the island observatory, perched atop a dormant volcano, where it was first spotted. In Hawai'ian, repeating a part of a word adds emphasis, and so 'Oumuamua's name described it perfectly—the scout from really, really far away.

By the time we saw it, 'Oumuamua's reconnaissance of the inner Solar System was complete and it was already heading away from the Sun without, as far as we could tell from the radio telescopes straining to listen in, ever emitting a signal. This first interstellar visitor was a silent one. As we watched it recede from us, more data was obtained. The object seemed to be a bizarre shape—elongated, like a cigar, and unlike anything the denizens of Earth had seen before. It seemed to be tumbling end over end as it followed its path around the Sun in a chaotic and unusual pattern. What could be responsible for this strange behavior and its bizarre shape? Was it out of control? Did some traumatic encounter long ago send it careening through the galaxy? Whatever was causing its behavior, 'Oumuamua was already heading out into the interstellar vastness from which it had come, speeding up as it did so.

Did this acceleration mean that it was firing its engines? Some thought so. Powerful radio telescopes redoubled their efforts to listen for signals coming from the departing visitor, hoping to eavesdrop if it was reporting back to its home base. No signals were found; perhaps they weren't there, or maybe they were simply too faint to be picked up. Perhaps whoever—or whatever—sent 'Oumuamua to visit us uses technology we can only dream of to pass messages among the stars, or perhaps, in a blow to our collective ego, there was simply nothing of interest around the Sun to report on. Just another yellow dwarf star, with a set of rocky planets gathered close to it and a few gas giants farther out. One inhabited world, with a primitive species beginning to take its first steps into space but still confined to orbits that cluster close around their own world. The changes we've wrought on the Earth's atmosphere show the state of our civilization: industrial, but not yet at the point where we've learned not to damage our planet. A problem for us, but not for the rest of the galaxy. The verdict, I imagine: "Mostly harmless," just as the *Hitchhiker's Guide to the Galaxy* famously had it.

An artist's impression of one possible model of 'Oumuamua, which looks rather cigar-like. In talks, I show both this and the actual telescope image, which shows only a point of light. We can infer a lot from watching a point of light move across the sky. *Credit: ESO/M. Kornmesser*

Everything I've written above about 'Oumuamua is true. In late 2017, astronomers poring through data from a survey telescope called Pan-STARRS* found something moving against the static background stars. Follow-up observations showed that it was on an orbit that, rather than being neatly bound to the Sun, as the Earth's is, was what astronomers call hyperbolic. This means that it is not permanently captured by the Sun. Instead, tracing its orbit shows 'Oumuamua arriving from beyond the Solar System and, after this one brief visit, that it will head back out into interstellar space, never to return. It was this unique orbit that first drew our attention to 'Oumuamua and led astronomers to point telescopes at it. The first confirmed interstellar object attracted an enormous amount of attention, which intensified as it was realized quite how weird 'Oumuamua really is.

Strictly speaking, it's the first interstellar object of any real size; we're not counting the interstellar dust that drifts through the Solar System all the time—or rather the clouds of dust through which we pass as the Sun orbits the center of our Milky Way Galaxy. These get caught up in meteorites, and a few grains have also been identified in samples collected from deep space by the Stardust probe, but they're not as impressive as a substantial asteroid.

Assume, for now, that 'Oumuamua is natural, and not a spaceship. It is still fascinating; this tiny rock carries with it

---

* It stands for PANoramic Survey Telescope And Rapid Response System, so I guess PSTARRS would be a more accurate name. Astronomers play with acronyms in the same way a karaoke singer in a bar late at night plays with notes—they understand there are rules, but they're surely not going to pay attention to them.

the story of its origins in a star system a long, long way from here. Its odd shape, its tumbling, and its weird acceleration are clues to its nature and history and can tell us about the conditions in which an alien Solar System must have formed. The fact that we've beaten the odds and found something this small and dark at all suggests that there could be many such objects flying through our own Solar System. The best estimates, taking into account the appearance of both 'Oumuamua and a second interstellar visitor, Borisov, which arrived a couple of years later (about which more shortly), suggest that there could be one such rock flying through the Solar System, within the orbit of Neptune, at all times. We just don't see them because they are so faint and move so quickly. If that's right, there are so many of these things traveling the galaxy that they may play an important role in the early stages of planet formation, and two of my colleagues, Susanne Pfalzner and Michele Bannister, believe that a visit by an object like 'Oumuamua could have been responsible for kick-starting the formation of the Earth itself.

We understand the basics of how planets form, with material left over from the process of star formation ending up in a rotating disc around the newly born star. In the disc there are tiny grains of dust, of silicon and carbon—think grains of sand, but smaller, maybe a tenth of the size of anything you'd pick up on the beach—and far from the star some of these will develop an icy coating of carbon monoxide or water ice. As they swirl around the star they collide and stick together, forming larger and larger particles. Eventually, clumps of material big enough for gravity to hold them together are formed, and such large bodies quickly accumulate more material from the disc they're swimming in. The largest of these proto-planets can even consume the gas in the disc, growing from a small seed perhaps the size of the Moon to something as big as Jupiter, by hoovering up hydrogen.

The only problem is that there's a stage in the middle where something the size of a boulder is made. Boulders are sufficiently small for gravity not to play much of a role keeping them together, but they are substantial enough that when you slam two of them together you get a rubble pile rather than growing a bigger single object. In other words, in smashing boulders together you go backward, producing smaller rocks, not larger ones. Understanding how planets grow to be larger than boulders—and, from what we can tell, do so both easily and quickly—is a big puzzle; but the arrival of an object like 'Oumuamua, already more sizeable than a mere boulder, could help solve it. If a passing interstellar object could be captured by the disc of material it could kick-start the process, being the seed that gets planet formation going. The Earth may have begun with the arrival of just such an object, and our planet may be built on alien foundations.

Based on the observation of only two interstellar objects, 'Oumuamua having been joined by Borisov just two years later, it is way too early to say whether this idea is correct, but 'Oumuamua's brief visit and the speculation that followed represent exactly the kind of discovery that made me fall in love with astronomy as a kid and which still thrills me today. The passage of a tiny rock that no one expected can upend all our ideas about such fundamental issues as how the Earth started to form.

So what do we really know about 'Oumuamua? Let's start at the beginning. The Pan-STARRS telescope that found it is situated on the summit of a Hawai'ian island; not the Big Island, where some of the world's leading observatories sit atop Mauna Kea, but the island of Maui, at the summit of Haleakalā. This was the first modern observatory in the islands, like its neighbors, placed high on a volcanic summit. The name, which translates as "home of the Sun," is especially appropriate for the site of the world's largest solar telescope, now under construction, but it is a great place for nighttime observing too, with more than three hundred crisp and clear nights every year.

The ability to observe for night after night, unmolested by cloud cover, is exactly what astronomers who track the sky's moving objects need. This is an old game—nineteenth-century astronomers found the asteroid belt by looking for objects whose movements stood out against the background stars. The first few such discoveries were celebrated as something special, and Ceres, the largest asteroid and the first to be discovered, was originally considered to be the Solar System's eighth planet. Pallas, the second, and Juno, the third, were also commonly referred to as newly located planets, albeit with less and less fanfare each time; but as the discoveries mounted and the number of planets in the Solar System threatened to grow alarmingly, something had to be done.

Ceres was relegated to minor planet status[*] and the word "asteroid" was invented (by William Herschel) to refer to these things. Asteroids were for a long while considered to be

---

[*] Similarly, we once considered Pluto special enough to be a planet, as it was discovered more than sixty years before any of its counterparts in the Kuiper Belt. Now, despite the continued rumblings of internet arguments, it is just one of many dwarf planets. Do not try to argue with me about this.

a nuisance, "celestial vermin" that cluttered up photographs of the more interesting, and usually more distant, objects that the astronomers were trying to study. Before too long, the asteroid catalog was growing, with official numbers assigned to each in turn once its orbit had been determined. By the second half of the last century, names and numbers were being assigned happily enough, the former having moved from the grand realm of Greek and occasionally Roman mythology to the firmly whimsical. Frank Zappa has an asteroid named after him (number 3834). So do each of the Beatles (4147 Lennon, 4148 McCartney, 4149 Harrison, 4150 Starr). So does Mr. Spock (2309), not the *Enterprise*'s science officer in *Star Trek* but a cat whose owner happened to be an asteroid-hunting astronomer.[*]

In all, just over twenty thousand asteroids now have names, and I have had the slightly surreal experience of peering into a telescope eyepiece and knowing that one of the "stars" in the field was no star at all but rather asteroid Lintott (4789).[**] (To get your own piece of celestial real estate—property rights not included—make sure to hang out with planetary scientists and drop reasonably unsubtle hints.) There are, though, more than a million objects numbered in the catalog, and thanks to projects such as Pan-STARRS that number is rising faster than ever—more than a hundred thousand asteroids are now added each year, most of which are likely to remain forever unnamed.

---

[*] The cat got its name as it had pointy ears and seemed logical, which I guess makes a certain amount of sense.

[**] I know two astronomers who use their asteroid's number as their PIN. Or at least they do until such time as they read this footnote.

It's the combination of digital cameras and clever data processing that's made the difference, together with telescopes designed with especially wide fields of view, so that they capture more of the sky with each shot. Pan-STARRS has the world's largest digital camera, a 1.5 gigapixel behemoth that spends each night tracking the sky. The volume of data is staggering; that in the last release was equivalent to a hundred thousand copies of the film *Armageddon*. Storing the latter would achieve little other than making sure that you could be really, really certain of watching Bruce Willis pretending to save the world from an incoming asteroid in a scientifically questionable way, whereas storing the former means that you have an excellent chance of finding the asteroid with our names on it in enough time for Bruce, or his less fictional counterparts, to do something about it.

Pan-STARRS, and telescopes like it, have accordingly done a good job of increasing the odds that we'll see the big one before it hits us. More than 90 percent of large near-Earth asteroids—those that cross our planet's orbit, and which could therefore one day collide with us—are now safely cataloged, and smaller objects are slowly being swept up. We can still be blindsided, and I'm writing this just a month after a 100-meter-wide object swept past the Earth at a distance of just 73,000 kilometers (much closer than the Moon!) without anyone noticing it was on its way, but the situation is much better than it once was.

The task of spotting asteroids is simply one of looking for, and then tracing, moving objects. A near-Earth object might move rapidly over the course of a single exposure made with a telescope's camera, leaving a streak on an image instead of a single point. More distant objects appear to move more slowly, and so require the careful comparison of images taken hours or even days apart, like some sort of cosmic "spot the difference." This technique, used to discover Pluto more than ninety years ago, is more difficult, and it's one of the reasons why the object that was to become known as 'Oumuamua wasn't seen as it entered the inner Solar System and swung around the Sun.

Only once it was on its way back out was 'Oumuamua identified by Pan-STARRS. On October 18, 2017, algorithms scanning new data, fresh from the telescope, alerted astronomers to the presence of an unexpected streak in that night's images. It was faint, and from what could be seen of its orbital track it appeared to move rather like a comet.

Comets are unstable creatures, often described as dirty snowballs. The repeated heating they suffer as they swing past the Sun again and again can destabilize them. On a small scale, that's what causes them to grow their beautiful long tails, which stretch away from the Sun as material sublimates into gas. The mercurial nature of the process means that it's never really possible to predict whether a given comet will be spectacular or a dud, and such activity has even been known to destroy a comet completely.

Typically, comets travel on a long, elliptical orbit that takes them out into the vast expanse of the outer Solar System, so that they spend relatively little time near the Sun. Comet Halley, for example, the most famous, last visited in 1986 and won't be back until 2061. I have a blurred memory of being taken out to look at it last time around—though we couldn't have seen much of it as we didn't have a telescope at home—and very much hope to catch it the next time, though it is forecast to be unimpressive. A much better view awaits those who can hang on until the twenty-second century, though, as Halley will be closer to Earth and thus appear much brighter when it arrives in 2134.

For most of the time, these wandering bodies live quiet and blameless lives. Getting near to the Sun is bad news for something that is mostly made out of ice, though, and as comets swing through perihelion—their closest approach to the Sun—their solid surface turns to gas, producing an atmosphere known as a coma, which surrounds the nucleus and, if enough dust and gas are liberated, also the long tail that marks a stereotypical comet. This can happen each time the snowy dirtballs complete an orbit; but most comets are usually most active on their first visit to the Solar System, when the ice is fresh.

Interstellar objects were expected to be comet-like. Just as it takes a lot of energy to power a rocket at takeoff so that it can climb away from the Earth, so launching things out of a Solar System requires energy to escape the gravitational pull of a star. Astro-navigators, who plot the course of spacecraft around the Solar System, consider the gravity well in which an object finds itself; climbing out of the well takes energy. It is therefore easier to escape from the outer Solar System than from the inner parts, as you're already most of the way out of the Sun's gravity well, and so we should expect objects that form around stars before being scattered out into the wider Milky Way to come from these more remote realms. In the inner regions of the early Solar System, the heat of the young Sun would have melted any ice. Rocky objects form here, while farther out icy bodies can take shape. In other words, if they come from the outer regions of their native Solar Systems, interstellar objects should be icy comets, not rocky asteroids. Our Sun has, we think, millions of comets orbiting it, almost all of them in a region called the Oort Cloud, which stretches nearly a light year away from the central star. Objects in these frigid outer regions of the Solar System are so far out that they should be much easier to disturb and liberate than any asteroid; indeed, as the comets occasionally collide or interact with each other, perhaps when stirred by the gravity of a passing star, then the Oort Cloud must slowly be evaporating, losing comets to interstellar space.

That's why we expected a comet. If it had been one, then 'Oumuamua, having never been near a star before, would have been covered with fresh ice and shown some sort of activity, developing not just a substantial coma but also growing a tail; yet it refused to match these expectations. As interest in its unusual trajectory increased and people used larger and larger telescopes to stare at it, no one could find any sign of comet-like activity. By the time, later in October, that its discovery and identification as an object on an interstellar path were announced via the telegram service of the International Astronomical Union,* even observations with the Very Large Telescope in Chile had failed to find anything and 'Oumuamua was thus classified as an asteroid. Specifically, a normal asteroid on a strange trajectory, but 'Oumuamua didn't stay normal for long. Continued observations allowed astronomers to trace its path and show that it

---

* Sadly, its telegrams are electronic rather than being delivered in uniform or on bicycle to the observatory door.

must have come from outside the Solar System. A new label was created, and it was cataloged as 1/I—the first interstellar body to be observed crossing our Solar System. Even now we don't know where it started its long journey, but we don't think it's from any of the stellar systems in the Sun's local neighborhood; this wanderer has found its way here from beyond our little corner of the Milky Way Galaxy, traveling for perhaps billions of years.

While debates about its origins, its nature, and its name were proceeding, observational astronomers kept their eye on this faint speck of light, trying to glean what they could. The fact that it was faint, the reason it wasn't spotted on the way into the Solar System, indicated that it was a small body, likely no more than a few tens of kilometers across. It was difficult to be more exact without knowing what 'Oumuamua was made of; the brightness of a body that shines by reflected sunlight depends on its size and on how reflective it is, a quantity called the albedo, so a shiny small object can be of the same brightness as a dull larger one.

Another mystery was soon added. Observations showed that 'Oumuamua was flickering, changing dramatically in brightness as it traveled through the Solar System. Sometimes it brightened almost ten times over the course of twenty-four hours. This is certainly unusual, but there is, we think, a simple explanation. 'Oumuamua is not round, and it tumbles as it proceeds along its path, so that in any given observation we see more or less of it (and the fraction of it that is in sunlight changes over time too), and so its brightness changes. This isn't entirely unexpected. Small bodies like the asteroids in our Solar System have all sorts of odd shapes: Gaspra, the first asteroid visited by a spacecraft, looks like a potato, for example, and the comet that the ESA's Rosetta spacecraft spent three years exploring looks like a large, if icy, rubber duck suitable for bath time. This distinctive shape proved unexpectedly useful when planning Rosetta's movements, as the team in mission control were able to refer to the head, body, or beak of the object. Even compared to a duck, 'Oumuamua is unusual by the standards of other small asteroids, its elongated shape, at least six times longer than it is wide, standing out.

We do know of a few Solar System objects shaped like this, but they are exceedingly rare. It reminded many people of the spacecraft in Arthur C. Clarke's remarkable story *Rendezvous with Rama*, which imagines an apparently abandoned cylindrical spacecraft entering the Solar System. In Clarke's novel, plucky astronauts make a close-up inspection of Rama and are able to confirm its artificial nature, though its origin, ultimate purpose, and the reason for its visit to our neighborhood remain obscure. Like 'Oumuamua, it leaves the Solar System, trailing mysteries behind it.

Preeminent among 'Oumuamua's mysteries was the way that it tumbled. It did not rotate cleanly, like a spinning American football, or like a majorette's baton thrown up into the air to spin end over end. It seemed to be more like a table tennis paddle that has been thrown into the air with a snap, so that any observer would see the handle and rubber surface of the paddle twist and turn as it falls. That snap suggests 'Oumuamua had a violent origin, perhaps as a fragment of a once larger body spun off after a collision. If so, then over the course of what might have been a journey billions of years long toward us, 'Oumuamua's tumbling motion had maintained the story of its beginnings, waiting for us to decipher them.

Alan Fitzsimmons of Queen's University Belfast (QUB) is one of the most cheerful people I know, an attitude reflected in his extremely loud Hawai'ian shirts. His enthusiasm for all things cometary is legendary. He is a regular on the BBC's *Sky at Night* TV program, which I present, though both of us once got into trouble with the producers for sounding rather *too* cheerful as we discussed the possibility that a giant asteroid might do for humanity what one did for the dinosaurs sixty-six million years ago. I always slightly distrust cheerfulness, and so when I visited QUB in 2018 to talk 'Oumuamua it wasn't too much of a surprise to find Alan, grinning manically, advancing toward me holding a flaming blowtorch. I just assumed he'd finally flipped.

As it turned out, Alan's blowtorch was used on a layer of sugar to demonstrate why, despite its lack of coma and tail, he and his collaborators were still claiming 'Oumuamua for Team Comet. Applying flame to the sugar produced a rather pleasant caramel smell and a crust of dark material on top that encased the unaffected sugar within, protecting it from the effects of a further assault with the blowtorch. The idea he was proposing is that long exposure to the interstellar environment of the Milky Way, and particularly to cosmic rays—which the Sun's magnetic field does a good job of shielding us from, but which flow freely through the rest of the galaxy—has "weathered" the surface of what once was a perfectly normal comet, producing a crust on top and keeping the fresh ice trapped inside.

A similar process involving cosmic rays is responsible for some aspects of the Moon's appearance. The regolith—lunar "soil"—is much brighter than the surface we see, but over billions of years its cumulative bombardment by particles in the solar wind has darkened it. It is certainly true that younger parts of the lunar surface often appear lighter; if you look at the full Moon, you can spot rays of fresher material spreading out from some of the more recent large craters. Over time, these will fade to match the color of the rest of the surface.

Spending time in space is bad for an object's complexion. In the case of our interstellar visitor, volatile material, such as ice, will have been lost from the surface in this weathering process, protecting 'Oumuamua from the Sun's heat. Somewhere in there might be a fresh icy core, but it's trapped inside a crispy shell of weathered material. Essentially, it will have developed its own heat shield, allowing it to pass through the inner Solar System unscathed. It's a neat idea—and the faintly red color that observations had revealed it to have (similar to that of bodies like Pluto and its moon Charon, which have spent billions of years hanging out in the outer Solar System, where they are relatively exposed to the cosmic rays that come from the rest of the galaxy) suggests that Alan and his blowtorch may be on to something.

For this idea to work, 'Oumuamua must have spent at least tens of millions of years en route, enough time for the slow weathering process to take place. We must imagine it traveling unimaginable distances only to pay the briefest of visits to our Solar System, and while it was here it had one more surprise in store for us. As it receded from the Sun in early 2018, it sped up.

Not by much, but enough to be noticeable by those watching from Earth. I'm used to thinking of the motion of the planets and every other celestial body as being dictated solely by gravity, so this seemed hard to explain. Gravity pulls at things, therefore the only significant force acting on 'Oumuamua, the Sun's gravitational attraction, should have been slowing it down, not speeding it up, as it headed out of our star's sphere of influence and climbed out of our gravity well.

The confounding thing is that such behavior is perfectly normal, not for asteroids but for comets that develop substantial activity. The heating of ice to produce a coma and tail can easily exert a considerable force on the main body. As it is warmed by the Sun, jets of material, often spectacular, can shoot up from a comet's surface, turning it into what is essentially a rocket. If 'Oumuamua was losing about a kilogram's worth of material to space every second, the observed acceleration would be explained. That isn't a huge amount, but it adds up; a rough calculation quickly showed that this would mean that its passage through the Solar System would have cost 'Oumuamua about 10 percent of its mass. A crash diet, then, but not completely unprecedented. But where was this lost material, and why hadn't we seen it in the form of the missing coma and tail?

This realization sent most of the planetary scientists I know who had been keeping an eye on the visitor back to check their observations again, bent on the urgent task of trying to reconcile the need for enough cometary activity to account for 'Oumuamua's motion with the apparent lack of anything that might resemble what a comet should be doing. An interesting puzzle, suitable for long stares into middle distance, arguments at the whiteboard, and perhaps a third coffee of the morning, but not, I would think, enough to overthrow everything we thought we knew about 'Oumuamua. Most people who have worked on the problem today think that you can explain the observed acceleration without too much trouble, especially given our lack of knowledge about what, precisely, 'Oumuamua is made of.

One man who disagrees is Avi Loeb, until 2020 the long-serving chair of the Harvard Astronomy Department. Avi is a well-respected cosmologist, a theorist of the early Universe with a reputation for thinking broadly about any problem he turns his mind to. His publication record is dotted with several papers that would make anyone's career, even some far from his home specialism; he was, for example, one of the first to suggest that we might find exoplanets by looking for the bending of light from distant stars, which encounters the gravity of an exoplanet en route to us. This is a method now happily put to use and providing fascinating results, including the discovery of a population of free-floating large worlds, wandering between the stars, that may even outnumber their counterparts in conventional orbits.

Some theorists are careful and miserly with their predictions, cautiously revealing their ideas only when every possible detail has been pinned down. Others—and I think Avi would be only too pleased to count himself among them—are happiest working in public, spinning out thoughts and ideas as quickly as possible in the hope of stimulating others into either considering the same ideas themselves or chasing down ways of testing his predictions using observations. This practice, of rapidly writing preliminary papers with new ideas, is great fun, and it can be rewarding for all involved.*

It wasn't that surprising when Avi turned his attention to 'Oumuamua. Avi and his collaborator and then-student Shmuel Bialy thought they'd cracked the problem of the visitor's behavior when they published their results in a paper entitled "Could Solar Acceleration Pressure Explain 'Oumuamua's Peculiar Acceleration?" We'll come to the specifics of their idea in a moment, but the first sign that this was no ordinary paper was in the last line of the abstract, which reads: "We discuss the possible origins of such an object. Our general results apply to any light probes designed for interstellar travel."

In the pages** of a sober scientific journal, they were seriously proposing the flight of fancy with which I began this chapter: 'Oumuamua, they argue, may have been an alien probe, designed—for some purpose—to fly through and survey and monitor the Solar System. Could 'Oumuamua be an alien spacecraft?*** Loeb chairs the advisory committee of Breakthrough Starshot, a companion program to Breakthrough Listen, a privately funded project to develop a fleet of probes the size of postage stamps to observe planets around neighboring stars. 'Oumuamua looks to him like a sail at least a hundred

---

* A recent paper of Avi's gave me a happy afternoon trying to work out whether his idea that activity around the Milky Way's central black hole could be responsible for a generation of rocky planets, for example, was at all plausible.
** Web pages, these days.
*** A good example of the rule about questions mentioned earlier.

meters across, a craft under active control: a larger and more sophisticated version of the system his team wants to design and build.

With the release of that one paper, coupled with Avi's status as "Harvard's top astronomer," the world wondered: "Have we found aliens at last?" So what was the evidence Loeb and Bialy presented? The idea set out in the paper was simple. Science-fiction writers, engineers, and dreamers of all stripes have long proposed the idea of a solar sail as a means of traveling across the galaxy. If you have a large enough sail, then the pressure of sunlight hitting it will slowly but steadily accelerate a spacecraft attached to it. The idea is elegant—who wouldn't want to glide across the galaxy on beams of pure light?—and cheap, because unlike any other form of propulsion you don't need to carry fuel with you. There are downsides, mostly because the initial acceleration is very slow, but high speeds can eventually be achieved if you're patient enough to just keep on accelerating. A few demonstration missions have shown that the technique works, at least on a small scale, in practice as well as in theory, and it's not ridiculous to imagine a fleet of solar sailcraft treading lonely skyways between the stars and receiving a boost from the light of each star system they visit.

This idea is, as I've said, an old one, but it was clearly revived now because of 'Oumuamua's strange shape. As I mentioned above, the data suggested that 'Oumuamua was six times as long as it was wide, but that's an estimate based on just a few weeks of observations. It's not impossible that the shape could be much more extreme, and so Avi and his collaborator proposed that 'Oumuamua was indeed a giant sail, no thicker than a centimeter and stretched out broadly for maximum speed between the stars. Their paper is actually reasonably entertaining; they calculate that any such sail launched into interstellar space will eventually be brought to a stop by collisions with interstellar dust particles, limiting the craft's range to the nearest thirty light years or so. There are plenty of stars within that range, several of which are already known to have planetary systems, any one of which could harbor a civilization capable of spying on us.

Yet beyond the obvious attraction of wanting to believe in aliens, there is no additional evidence in our observations for an extreme shape. It's not a better explanation for what we see than the plain old cigar configuration, which is relatively useless for sailing, and maintaining that the object is an effective sail makes 'Oumuamua seem even odder than it needs to be. In science, it's not a bad rule of thumb that the less weird an explanation is, the more likely it is to be right.

What about the idea that 'Oumuamua may have been intentionally aimed at us? The only evidence for choosing this scenario is that objects like 'Oumuamua were, before its arrival and discovery, expected to be rare, and yet we've found one. Its unexpected presence, if you're already thinking aliens, becomes clear evidence that someone—or something—out there is targeting us.

Fine, I suppose. I am even an editor of the journal that accepted the paper—though I had nothing to do with it—and I'm guessing that the referee found nothing wrong in the calculations it includes. This sort of intellectual game of "what if" is played in physics and astronomy departments around the globe whenever anything unusual crops up. But most of us leave our speculations to the coffee-room whiteboards rather than writing up claims about alien spacecraft and sending them around the world. Avi is made of sterner stuff and would go on to expand his ideas in a book, which essentially claims that the combination of odd properties that 'Oumuamua possesses is so strange that they simply cannot be natural. The tumbling, its weird acceleration, its presence in the Solar System, and the fact that it seemed, on the way in, to be hardly moving relative to our local collection of stars to him mean that the odds of it being natural are a trillion to one.

Avi's book also contains a reading list of over two hundred separate works, every single one of them with Avi as a coauthor. This is not the work of someone closely collaborating with the experts who observe the small bodies of the Solar System at all times, but someone plowing a lonely furrow. The idea of 'Oumuamua as an alien spacecraft has, to my mind, if not to Avi's, been comprehensively debunked. Of the arguments ranged against the alien hypothesis, I find most convincing the fact that if the thing really is a solar sail, it would orient itself so that the sunlight would hit it square on, producing the maximum acceleration. That means it would change its angle as it moved as viewed from the Earth. But if it was as extreme a shape as a pancake, as Avi proposed, for the solar sail theory to work it should have changed brightness even more rapidly and dramatically than we saw. It didn't, so case closed.

As for the rest, from the perspective of observational astronomers, the oddest aspect of 'Oumuamua is that we know so much about such a tiny thing: few bodies of similar mass in our own

Solar System have been considered significant enough to study in detail. The world's most powerful telescopes were pointed at 'Oumuamua because of its unusual trajectory. A garden-variety comet on a normal orbit would never get such scrutiny from overworked and in-demand facilities.

That is, I know, not an argument that convinces Avi, who has moved on to ever-more-creative pursuits; when I last saw him, he asked me if I knew people who might finance a trip to the Southern Pacific, where he believes a possible alien artifact, disguised as an interstellar meteorite, has fallen. That trip did take place, though no one who has studied meteoritics (or the seafloor) thinks he could possibly have found anything of interest.* He has also placed microphones on the roof of Harvard's astronomy building, listening for any alien activity that might reveal itself through sound.

Meanwhile, while we were all still debating what 'Oumuamua might or might not be, the world's astronomers were again surprised by the arrival of a second interstellar visitor. This one was brighter, discovered while it was still on its way into the Solar System by the remarkable amateur astronomer Gennadiy Borisov in Crimea. Borisov's day job is maintaining the instruments at the Crimean Astrophysical Observatory, but at nighttime he observes with his own handmade instruments, and it was one of these that led to his find in August 2019. Spotting a faint, fuzzy blob in his images, he thought he'd found a

---

* Apart from anything else, any debris from the particular meteor he's interested in will be mixed with all the others that have fallen there over the last few thousand years.

comet, and it was named after him, as comets—in contrast to asteroids, which must be named for other people—have always been named for their discoverers. From the little he could glean about its orbit, he also thought it was heading straight for Earth, and immediately reported it on the web page where astronomers keep track of dangerous asteroids.

This caught people's attention, and by the time September came around enough new observations had been gathered to refine our knowledge of its orbit, showing that the Earth was safe and that Borisov, like 'Oumuamua before it, had come from beyond the Solar System. This new visitor was a little larger than its predecessor, and it lost plenty of material as its icy surface was exposed to the Sun's heat. It was indeed what those few theorists who had thought about interstellar objects had predicted before 'Oumuamua confused everyone: a comet. It didn't do anything weird that could indicate alien activity, and its discovery suggests that there is a whole population of these things out there, waiting for us to discover them.

Soon we will have surveys that should locate a handful of interstellar objects each year, and we'll find out exactly how odd the first one was. We might even get a close-up look: the ESA, collaborating with the Japanese space agency JAXA, is planning a mission called Comet Interceptor, which will be launched into orbit and will then just wait, being activated only by the close approach of a comet from the outer Solar System. So far we have only visited comets on predicted orbits, as we need time to plan, build, and launch a probe to catch them. Instead, Interceptor will be ready and waiting, and it's possible that it will be sent off not to a normal comet but to an approaching interstellar object. I hope so. These visiting objects, which tell us about the process of planet formation in systems we will never visit ourselves, and which bring a tangible piece of the Milky Way to our neighborhood for a while, are exciting enough without the need for aliens at the steering wheel.

Chapter 4

CELESTIAL VERMIN

As a kid peering through my small telescope, dodging street-lights outside our house to try to cover the sky, I dreamed of finding a comet wandering the skies and catching it heading for the inner Solar System. It was easy for me to imagine the world's wonder as they witnessed Comet Lintott blazoned across a twilight sky, its magnificent twin tails reaching up from the horizon.* As I looked up, my head would fill with images I'd seen of the great comet of 1843, with a long, straight tail that seemed to stretch right around the sky, or Donati's Comet of 1858, with its braided tail spreading feather-like across the northern sky.

Let's be clear. In attempting to find one, I was completely out of my depth. In the early 1990s there were still a handful

---

*   I also think a chemical element would be fun. Lintottium sounds great.

of visual comet hunters around. I remember meeting the marvelous George Alcock, who, from his base outside Peterborough, England, scanned the skies with binoculars, comparing what he saw each night with the star patterns stored in his head, looking for anything out of place. His memory and persistence for decades enabled him to discover five comets, but by the time I got anywhere near an eyepiece the era of automatic sky surveys had started. Robotic cameras now regularly patrol the sky, powered by software that is alert to the presence of any moving object. Some inspection by human eyes of the data flowing from these systems is still needed to weed out the real discoveries from camera noise and other artifacts, but the process is now largely automated and involves looking at likely candidates on a computer screen instead of spending long, chilly nights under the stars.

Though these modern methods give us many more comets, most fail by some distance to match their most famous predecessors. Predicting cometary behavior is notoriously difficult, as the brightness and prominence of a comet will depend on how the icy nucleus changes as it approaches the Sun. As comet nuclei seem to have a variety of compositions and shapes, and the effect of heating them is somewhat random—a process more like a glacier calving icebergs in the Antarctic summer than the even melting of an ice cube on a picnic table on a hot day—at best we can usually give a rough guess as to the odds of a spectacular display. Anticipating being disappointed is likely a winning bet; the passage of most comets through the inner Solar System is unremarkable, though a few become bright enough to be seen with the naked eye as faint, fuzzy blobs, moving from night to night against the background stars. As our surveys become more sensitive and comprehensive, catching fainter and more distant comets, the odds of any particular discovery being remarkable drop. Even were I to join the ranks of astronomers filtering data from automated telescopes, any putative Comet Lintott would likely be a dud.*

There are glorious exceptions, of course. I remember fondly Comet Hyakutake, which in 1996 displayed a tail that stretched nearly from one horizon to another, and Hale-Bopp, a year later, which sat picture-perfect above the northern horizon for weeks, easily noticeable even by those who normally pay no attention to the sky above their heads. Nothing like them has since been seen in the Northern Hemisphere.** Watching Earth's brief encounter with these two objects ranks among the most memorable experiences of my observing life, but it was a much fainter comet just a few years earlier that first taught me that space is a place where dramatic and unexpected things happen.

Comet Shoemaker-Levy 9 (SL9) was first discovered by a legendary team of three astronomers. Eugene Shoemaker taught geology to the *Apollo* astronauts, helping what was mostly a band of high-achieving pilots and military men get ready for the scientific exploration of the lunar surface. His wife, Carolyn, was an observer of renown, with the keenest of eyesight, and held the

---

* If the attraction of finding a comet, no matter how unimpressive, remains, then my best advice is to do a PhD with Alan Fitzsimmons at Queen's University Belfast, where students take shifts sorting through possible Solar System objects found in data supplied by the Zwicky Transient Facility, located on Mount Palomar, in California.

** I'm hoping that writing this sentence in the middle of 2023 will guarantee that you, dear reader, can enjoy a glorious string of comets that now must surely arrive between submitting this text and publication.

record for most cometary discoveries for many years. The third member of the trio, David Levy, was another fine observer and a science writer of some repute. Between them, the three friends found nearly two dozen comets, but their ninth—SL9—was special. Soon after its discovery, images revealed that this new comet did not have a single nucleus but several, which stretched in a line across a telescope's field of view like a string of pearls. Nothing like it had ever been seen before, but an analysis of the comet's orbit provided a simple explanation. A close encounter with Jupiter meant that the previously intact nucleus had been ripped apart by the giant planet's gravity, producing a string of cometlets.

The best part of the SL9 show was still to come. The encounter had also resulted in the capture of the comet by Jupiter, so that it orbited the planet as a temporary moon[*] rather than continuing on its journey around the Sun, and the cometlets were due to collide with the planet on their next close approach. As the fateful day neared, the Hubble Space Telescope produced spectacular pictures of the comet as the fallout from its catastrophic encounter with Jupiter continued. Fragments vanished or broke up on a daily basis as they fell toward the giant planet's atmosphere. The timing of the impact of those pieces that remained could be easily predicted, as could the impact site: in Jupiter's southern hemisphere, frustratingly just on the side of the planet facing away from Earth and a planet-load of eagerly waiting observers. Luckily, the Galileo spacecraft, then en route to begin its tour of the Jovian system, did have a direct, if distant,

---

[*]   Or, I suppose, as a string of moonlets.

view, but the rest of us knew that we would have to wait for an hour or two after each cometlet plunged to its doom for Jupiter's rotation to bring the entry site into view.

Despite this, anticipation ahead of the first impact was immense. At no time before or since had so many of Earth's telescopes, large and small, been pointed at a single target. It was July, and as the summer twilight deepened and Jupiter appeared, shining brilliantly, among the battery of instruments arrayed with hope (if not expectation) of seeing something was a small reflector with a six-inch mirror, plonked on the patio in a South Devon backyard, which I'd positioned carefully so that the planet stayed just above the neighbor's hedge.* No one knew what the effect of a piece of comet a kilometer across plunging into Jupiter's thick atmosphere would be. The comet would certainly burn up as it fell, and it later turned out that the resulting plume of hot material could be seen in images from Galileo's infrared camera, but most believed that the vastly more massive Jupiter would be unaffected by such an event. After all, the gas giant offered no solid surface for the comet to hit.

Yet we were all surprised. Jupiter, even through my back-yard telescope, was noticeably bruised: a new spot, blacker than any I'd ever seen, marked the impact site. I will never forget the shock of staring through the eyepiece at something so completely unexpected. There was a moment of absolute surprise, disbelief that made a second look essential, and then a surge of elation that something new, something different, was visible. There was

---

*   Not a consideration for most professional observatories, or the Hubble Space Telescope.

nothing unique about my perspective, and of course many others got there ahead of me, or had bigger telescopes or better views, or could record what they were seeing with images, but that one moment felt to me like a personal encounter with a changing cosmos.

It turned out that the impacts had been much more dramatic than anyone expected. Those plumes, seen by Galileo, reached thousands of kilometers above Jupiter's cloudtops. The atmosphere, where the comet fragments hit, was heated to several tens of thousands of degrees centigrade. And over the next few days, as pieces of the comet rained down on Jupiter, a string of these dark marks appeared, one by one, to my increasing wonder and excitement. Over the next few months, we watched as the bruises changed, faded, and eventually merged into a new belt spanning the planet, which then slowly disappeared as the churning of the Jovian atmosphere erased all evidence of the impact. In these bruises we had collectively witnessed for the first time the effect of a major impact on another Solar System body, something not seen in centuries of scrutiny since the invention of the telescope. (A postimpact search of the archives revealed a few instances of black spots appearing on both Jupiter and Saturn that look suspiciously like those seen after SL9. Never before, though, had an impact been anticipated and observed in real time.) In the aftermath, it was hard not to think about what such an impact would have done to the Earth. Jupiter's greater mass and subsequently stronger gravitational pull make it by far the most likely target for asteroids or comets on a collision course but impacts with Earth do happen on an uncomfortably frequent basis.

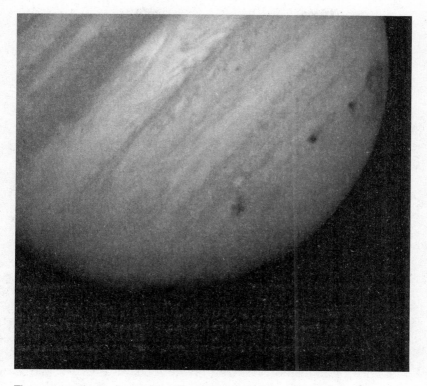

The spectacular bruises left by Shoemaker-Levy 9 in Jupiter's atmosphere, as seen by the Hubble Space Telescope. The smallest things in this image are two hundred kilometers across, and the scars are roughly the size of the Earth. *Credit: Hubble Space Telescope Comet Team and NASA*

More than fifteen thousand kilograms of celestial material ends up reaching the Earth's surface each year, though most of it falls unnoticed over oceans or deserts, and mostly in fragments too tiny to trouble anyone should it land elsewhere. Larger pieces do reach the surface; the most recent fall in Britain, on February 28, 2021, delivered a smattering of meteorites to the small Gloucestershire town of Winchcombe. The Wilcox family, watching Sunday evening television, actually heard the fall, noticing an odd sound as a piece of meteorite splatted into the ground in front of their house. Finding what looked to be the remains of a crumbled charcoal briquette on the drive the next morning, they'd have been forgiven for tidying up. Luckily, the incoming meteor had been tracked by the UK Fireball Network, who make use of cameras pointed skyward by volunteers across the country to plot the trajectory of such objects, and the local news had carried a request to contact London's Natural History Museum (NHM) should anything unusual appear.

The Wilcoxes carefully stowed their driveway meteorite in a (clean) Waitrose tote bag and called the museum. Ashley King from the museum's Earth Sciences Department, dispatched at speed, quickly identified what he had in front of him as the first fragment of the fall to be found, but he had arrived too late to get it safely back to London that evening. He recalls a sleepless night in a budget hotel, sitting on the bed staring at a plastic bag full of brand-new, fresh meteorite sitting next to the TV, the scientific find of a lifetime for someone who studies such objects. As the sample was conveyed carefully back to South Kensington the next morning, heading rapidly in the other direction was a small but dedicated team of meteoriticists from across the country, accompanied by a couple of family members and friends who got caught up in the excitement and were pressed into service. Together, over the next week or so, this ragtag team carried out a fingertip search of surrounding fields and footpaths, hunting for more meteorites.

If you carefully cross Gloucestershire fields looking for small, black objects that stand out against the grass, you mostly find sheep droppings.* After nearly a week the team hit paydirt with the discovery of a large piece of Winchcombe meteorite by a team from the University of Glasgow, accompanied by Mira Ihasz.** Mira isn't a meteor scientist, having joined the hunt as the partner of the coordinator of the Meteor Alliance, but she made the most valuable discovery of all.

Crucially, her piece and the others recovered from Winchcombe and its surroundings were all found before it had rained. Rain would have disrupted the meteorite's chemistry, an important part of its scientific interest. Winchcombe is especially exciting because it turned out to be a rare carbonaceous chondrite, a member of a class of meteorites that represent some of the oldest material in the Solar System. What had fallen on southern England that February evening was some of the oldest material that exists anywhere, a leftover building block from the time more than four billion years ago when the Earth and other planets were assembling. Only a hundred or so meteorites of this class are known, and few have been collected with such (Waitrose bag and all) alacrity and care as the Gloucestershire specimens; in the weeks after the fall, as the Natural History Museum scientists came to realize what they had, comparisons were made with

---

* There is a rich literature categorizing such "meteowrongs." My favorite story belongs to the NHM's Sara Russell, who was part of a group in Antarctica who recovered, meticulously cataloged, and sent off for analysis what NASA's Jet Propulsion Laboratory revealed to be a piece of chocolate left behind by a previous expedition.

** Meteorites are named after the nearest established settlement, hence the name now officially bestowed on the fall. By convention, if not the letter of the law, it's the nearest post office that counts, something that works well in the UK and less well in, for example, Antarctica.

the samples of asteroid Ryugu collected by the Japanese Hayabusa2 probe, which had to fly to the asteroid and back instead of simply waiting for a suitable sample to fall to Earth.

The chance fall of the Winchcombe meteorite offers us the opportunity to address one of the fundamental questions of modern science: Was the formation of the Earth, or a planet something like ours, the inevitable consequence of the physics and chemistry that operate around stars like our Sun while they are still forming, or was our planet and our subsequent history on it nothing more than a lucky chance born from chaos? It also provides a warning. Although it did little more than damage a driveway,* the meteoroid that led to the Winchcombe fall was completely unknown before it hit the top of the atmosphere and serves as a potent reminder of how vulnerable we are to the unexpected. Indeed, no detection system we can realistically imagine building would spot objects like this, which are perhaps just a few meters wide; they are simply too small, too faint, and too numerous for us to build up a useful catalog. Their small masses also mean that nonlinear effects, such as the results of solar heating on their rotating surfaces, have a significant influence on their orbits, so if we find and catalog such an object today there's no guarantee that it will be in the same place in a decade or so.

Larger objects exist and do, on occasion, hit the Earth. Perhaps the last big thing to do so crashed down on Tunguska

---

* Damage by a meteorite can be a lucrative affair: in February 2022 a dog kennel hit by a meteorite in 2019 sold for $44,000, double the price reached by the impactor itself. Cars hit by meteorites have also been known to gain in value; the same is presumably not true for people, though only a handful of cases, all mercifully nonfatal, of humans or animals being struck are well documented.

in Siberia in 1908, flattening trees over an area of more than two thousand square kilometers, though in such a remote place there was little else for it to damage. (The exact nature of what fell is disputed, as some think the pattern left by the blast may indicate a comet that disintegrated completely before reaching the ground, creating an airburst whose effects felled trees, rather than the traditional meteorite. No meteorite has ever been found buried in the soil.) Nowadays, there is more hope that we might be able to spot anything big enough to seriously damage a city should it score a direct hit well before any collision happens. Scanning the sky to find such near-Earth objects (NEOs), those whose orbits cross that of Earth, is something of a specialist task, but it is one being undertaken with seriousness by small groups of astronomers who talk earnestly about saving our planet.

Because such NEOs are almost always detected as they pass near Earth, they are not especially faint, bringing them within range of large amateur telescopes. The tricky part is to spot the interloper in rapid motion by comparing its position from hour to hour against the background stars, and to have the savviness to identify it for what it is, distinguishing a true NEO from a satellite or more distant denizen of the asteroid belt. Among the sharpest players of this game are a group of amateur astronomers who run a set of robotic telescopes at the Northolt Branch Observatories in West London, a light-polluted site that could lay serious claim to being the place with the worst possible conditions from which to observe the sky. Since 2015 the team at Northolt have been carrying out a systematic search for NEOs, reporting discoveries to the international clearinghouse for new

Solar System discoveries, the Minor Planet Center (MPC) in Cambridge, Massachusetts.

Take, for example, the object they found in observations made between April 10 and 12, 2020. After their announcement of the new find, it was "precovered"—detected in images taken before the discovery, but not noticed before—in surveillance from telescopes in Hawai'i and California. These provided enough data for the MPC to calculate a preliminary orbit for it and give it a formal catalog designation—2020 GL2. They announced its presence to the world, and here 2020 GL2 passed from being mundane to being something remarkable and potentially threatening. Shortly after news of the discovery spread worldwide, computers at NASA's Jet Propulsion Laboratory in Pasadena, California, warned that there was a one in four hundred thousand chance of a collision with the Earth just eight years later, in 2028.

That isn't a large chance. It is, in fact, the probability that your worst work colleague will get spontaneously incinerated by lightning in the next year,* but even such seemingly remote odds immediately made GL2 the most dangerous object in the books, not least because the observations in hand were not sufficient to allow a proper estimate of its size. If something big enough to change the global climate is heading our way, and due to hit us in less than a decade, then one in a few hundred thousand is far too close for comfort.

---

* It's also the odds of you suffering that grizzly fate, but our brains do funny things to probabilities as applied to ourselves, so best pick on Jeremy or Jenny or whomever instead. They're probably hoping you get hit anyway.

The immediate remedy was obvious. More observations were needed. These, taken over many nights by a variety of telescopes scared into action by the possible collision, were designed to reduce the uncertainty about GL2's orbit. The situation is rather like what happens when you watch a soccer match on TV and someone hits a shot toward the goal from outside the box. Hit pause just a second after the ball has left their boot, and it will be very difficult to predict its trajectory.* Wait a little longer before pausing and, thanks to the extra information we gain as we watch the ball move, we have a better chance of forecasting the outcome. We may still be deceived by some late swing, but essentially the longer we watch the ball the more information we have about its path.

GL2, as it turns out, required no new observations to quell any growing sense of alarm. Before anyone else could observe it, or at least before they could report their observations to the MPC, it became clear that the new object was no Earth-bound asteroid, threatening civilization, but was, in fact, ESA's BepiColombo spacecraft, flying past Earth as part of a series of maneuvers intended to send it to Mercury. The celestial geometry of its mission means that a direct route from here to the innermost planet is not possible, and Bepi has to make several fly-pasts of the Earth and Venus to adjust its trajectory between launch from French Guiana back in 2018 and ahead of its arrival in Mercury's orbit in 2025.

---

* Unless it's a Torquay United FC player, in which case I can tell you from long experience and with certainty the ball is going miles and miles over the bar.

This particular case closed with a rare formal MPC retraction, GL2 being deleted forever from its database, but it does illustrate the scale of the challenge in cataloging and identifying such objects. Confusing asteroids with spacecraft happens reasonably often; the one time I visited the MPC in person decades ago, a polite chat with the director was interrupted by the announcement from a colleague that an object whose trajectory was puzzling had resolved the crisis by firing its engines and moving off. ESA's Gaia and Rosetta spacecraft have also both briefly been allocated official asteroid numbers.[*] False alarms aside, the size of the catalog is enormous: twenty thousand asteroids that cross Earth's orbit are now known, and more are being discovered each week. On top of that, the majority of observations of moving objects caught in astronomical surveys submitted to the MPC seem not to relate to any known ones; these are likely to be new asteroids. Unless the observations cover many nights, they are usually insufficient to properly define the orbit of anything new and allow it to be added to the catalog. Such records of unidentified asteroids are filed under the slightly unfortunate title "one night stand," to dwell in obscurity until a future discovery makes sense of them, essentially by quickly pinning down an orbit. Progress is being made, though. Between the work of ground-based surveys and clever reuse of assets like NEOWISE, a repurposed infrared telescope still in space and now functioning primarily

---

[*] I'm not sure why it always seems to be European and not American probes that are picked up. It's hard not to conclude that the computers at the Jet Propulsion Laboratory are better able to track craft carrying the Stars and Stripes.

as an asteroid seeker, our list of the largest NEOs, those capable of really altering the Earth's climate and affecting the odds of humanity surviving, is now reckoned to be complete. It's only the smaller asteroids that could take out a city or a continent that we still need to worry about.

As the list of known objects grows, a group of SETI enthusiasts are paying attention, planning to search the catalog for their own purposes, hoping to find not only spacecraft but perhaps also an alien interloper, revealed by the oddness of its motion or orbit or by its color. The technique can be tested by trying to find Earthly spacecraft; the Tesla launched into orbit by Elon Musk as a publicity stunt, for example, is easily identified as something unusual on a plot of the colors of everything in the Solar System. Had it been an alien vehicle, it would have stood out; and just as 'Oumuamua revealed itself to be something interesting because of its orbit, the hope is that as we get better at detecting our own technology, we might find other explorers out there, spotting them as they pass through.*

No alien spacecraft have turned up yet, but the MPC files do furnish a long list of credible-looking targets that we should worry about. If we are to go the way of the dinosaurs, it seems likely that we will at least see our nemesis coming. Some efforts are under way to rate known objects by the threat that they pose, on the equivalent of the Richter scale for earthquakes or the Beaufort scale for wind, though based on the potential for disaster rather than on its effects. The best known is the Torino scale,

---

* As I write this, a summer student, Brian Rogers, is working on new techniques for identifying alien spacecraft. I'd keep an eye on Brian if you expect aliens to visit soon.

the details of which were mapped out at a conference in Turin.[*] The scale runs from zero—an object with next to no chance of hitting the Earth—to ten, which is a certain impact with the potential to cause a global disaster.

It is not unusual for new objects to rate a one on the Torino scale, though so far all of them have dropped down to zero soon enough, usually within a couple of weeks of discovery as more observations flood in and their orbits become better known. Whatever caused the Tunguska impact might have scored an eight; were such a blast to occur in a populated area, it would cause damage approaching that of a well-targeted nuclear bomb. The good news, though, is that no asteroid we know of at time of writing rates even a one on the scale, and there are no large asteroids (bigger than a few hundred meters) that have even a one in a thousand chance of hitting the Earth within the next few centuries. Those of nervous—or pessimistic—disposition can check the current list here: https://cneos.jpl.nasa.gov/sentry/.

We are, though, playing the cosmic lottery just by existing on this fragile planet. At some point our luck will run out, and the inevitable major impact will occur. How ready are we for this most dramatic form of cosmic surprise? One answer might be to contemplate another form of risk from space that we have been aware of for longer, in the form of solar weather.

Astronomers have long known that our Sun is not the completely stable, reassuring presence it appears on a balmy spring day. Rather than being the cheerful yellow disc that shines from

---

[*] If you're going to gather friends and colleagues to think about the end of the world, I have a certain sympathy with doing it somewhere beautiful with great Italian food.

the sky of children's drawings, it is a superheated maelstrom of plasma, powered by nuclear reactions that take place in a core at temperatures of trillions of degrees. The energy released by these reactions, which convert hydrogen to helium, creates massive convection cells as material low in the Sun's atmosphere is heated and rises, with colder material from near the surface sinking in turn to take its place. Allow a pan of thick soup on the stove to reach a vigorous boil, and in a similar process, bubbles will appear and splatter across your kitchen counter, though the situation in the Sun is made more complicated by the presence of our star's strong magnetic field.

The magnetic field is inevitably twisted as the Sun spins on its axis, faster at its equator than at the poles. This differential rotation was first observed by tracking the motion of the dark sunspots that appear on the Solar surface, themselves the product of the machinations and complexity of the magnetic field shaped by our star's rapidly moving plasma. The twisted magnetic field stores energy, and that energy has, sooner or later, to be released. When it is, in the form of either a solar flare or a more dramatic event known as a coronal mass ejection, substantial material can be flung into space. My friend and collaborator Chris Scott likes to describe these events, which he studies with cameras on board the twin STEREO spacecraft, built to monitor the environment between the Sun and Earth, as a billion tons of matter moving toward us at over a million kilometers an hour.

STEREO and a fleet of other spacecraft are seeking to understand this so-called solar weather and to give us advance warning of the arrival of a storm, because such events can have a significant impact on Earth and on our technology. Fortunately, we are somewhat protected by our planet's magnetic field, which deflects the worst of the incoming storm away from the surface, while channeling incoming particles to produce the magnificent light shows we know as the Northern or Southern Lights. Astronauts straying farther afield may not benefit from such protection. Keeping a weather eye on the Sun will be vital for anyone visiting the Moon; one of the largest and most powerful series of solar flares ever recorded took place in August 1972, between the return of *Apollo 16* from the lunar highlands and the landing of Eugene Cernan and geologist-astronaut Harrison Schmitt on *Apollo 17*. Had any of the crew been caught on the lunar surface during this storm, severe radiation sickness, if not worse, would have been a certainty, and even inside the lunar lander, protected only by their craft's thin aluminum shell, the astronauts may not have been safe. For longer voyages farther afield, the problem worsens. The journey to Mars is difficult to comprehend, as astronauts would spend many months exposed to any passing flare. Things aren't much better once you arrive. As the red planet has lost any magnetic field it once had, future settlers would be driven to an existence underground, so as to be shielded from the worst of the solar weather by the planet's natural rock, hiding below the surface to compensate for the lack of protection that the Earth's magnetic field provides naturally.

Along with working out how to pay for the trip, radiation is probably the biggest unsolved problem that prevents us from traveling far beyond the Earth's friendly confines. Still, storms can affect us down here too. The most famous and powerful flare ever observed took place in September 1859. Two British astronomers independently happened to be looking in the right direction and saw the bright flash of the flare itself on the solar surface. The event tends to be known as the Carrington Event after Richard Carrington, who connected what he'd seen on the Sun with the strange events on the Earth a few days later. The flash of light Carrington and Richard Hodgson had seen arrived at Earth at the speed of light, taking eight minutes to get here from the Sun's surface, but the particles released in the storm took nearly a day to reach us. This is, in fact, surprisingly rapid, a sign of the storm's power. Normally, events on the Sun affect the Earth's space weather a couple of days later.

When the particle storm arrived, telegraph systems responsible for carrying the bulk of urgent communication failed in both Europe and North America, and operators reported sparking boards in front of them as the system somehow absorbed the charge coming from outer space. Many took the understandable course of unplugging their systems from any power supply, only to find they could continue to send and receive messages regardless, due to the power of the insurgent storm. A newly electric society learned a lesson about the fragility of its systems in a world vulnerable to interference by the Sun; but as nothing nearly so serious as the Carrington Event has been seen in the more than 150 years since, we have gone on to develop an infrastructure that is still more important, and therefore still more vulnerable. Nearly a decade ago, a study by the insurer Lloyd's of London suggested that a repeat of 1859 would cost trillions today, a value arrived at by considering the damage to modern satellites and all the systems that depend on them. In March 1989 a lesser storm caused patchy blackouts of satellite communications, affecting users all over the planet, and overwhelmed the electricity system in Quebec. The power failure came quickly as the front of the storm struck, taking the systems from normality to offline in around a minute and a half; the lesson is that when solar storms hit, they hit fast. Without monitoring, we are inevitably blind to what is about to happen, just as the lack of astronomically qualified dinosaurs cost T. rex dearly.

In the last few decades, we have started to take monitoring space weather seriously. The UK Met Office now forecasts solar storms, based on observations of activity on the Sun, alongside their predictions of terrestrial weather. As I write, their website tells me that solar activity has been "low" in the last twenty-four hours, with only two flares and no more than five sunspot regions currently visible on the disc. Those sunspot groups seem stable—a good sign, if we are trying to avoid flares—and the observed solar wind is within normal parameters. It is reassuring, even if I haven't yet reached the point where checking the solar weather is something I do with my morning toast in order to plan for the day ahead.

If I ran a major electricity company or a satellite constellation, I suspect things might be different. Simple mitigating actions, like putting satellites in safe mode as a solar storm sweeps through, or preparing the electricity grid for an influx of particles, can make a big difference in the effects of solar weather. Even a few hours' warning of something heading our way would be enough to avoid most of the economic cost of a big storm. Over 150 years after the Carrington Event, we are, thanks to the deployment of satellites monitoring the Earth's atmosphere and organizations like the Met Office, at the point where a Carrington Event should be scientifically fascinating but, on Earth at least, nothing to trouble the stock market.

It took more than a century to go from Carrington's discoveries to routinely predicting and adjusting for the solar weather. Although we have known since the dredging of the seafloor in the late nineteenth century revealed unexpected rocks that celestial material was raining down on Earth, it was only in the 1960s that features such as Meteor Crater in Arizona were recognized as illustrations of the damage an impact could cause. (Gene Shoemaker, of SL9 fame, was one of the people responsible for developing convincing evidence that this well-known feature did in fact have a celestial origin.) Popular awareness of what a meteorite could do received a further boost in the early 1980s when physicist Luis Alvarez, his son Walter, and a few chemists discovered evidence that an impact with global effects had marked the change visible at the boundary between the Cretaceous and Paleogene periods five million years ago. The idea that the most recent sudden mass extinction (excluding the one that we're inflicting on the world right now) had a cause that came from outer space certainly concentrated minds.

We are amid a revolution in how we think about the threat from asteroids and meteoroids. Until the last few years, the focus has been on understanding the odds of detecting large rocks that might hit the Earth. Until now, though, we haven't had a solution to what we would do if we *did* find that something threatening was en route; it isn't as if we could just turn off civilization, hide underground, and wait until the danger has passed, like so many Martian colonists facing a solar storm. The logic has always been that the earlier a discovery of a potential impactor can be made the better, buying us time to deploy Bruce Willis or come up with a plan that would actually work.

A favorite suggestion is that, given enough time, simply sending a robot decorator to paint one side of the incoming asteroid white should be enough to deflect it. Such a mission would need to be carefully designed, but it seems like the physics is sound. Light from the Sun exerts a small but measurable pressure, which depends on the color of the body it shines on. If you can alter an asteroid's color early enough, the tiny differences in momentum that result lead to a change in orbit big enough to ensure the Earth is missed. This should be effective, but it doesn't make for fantastic Hollywood plots, and it does require a long period between discovery and the projected impact in order to be effective. To my disappointment, the most recent studies seem to suggest that the change in orbit would be so slight that we may never be able to give ourselves enough notice, rendering this rather lovely, low-intervention route to preservation impractical.

It might be wise to pay more attention to problematic asteroids in advance. I've been following the OSIRIS-REx mission closely as it visited asteroid Bennu, one of the larger bodies that has the potential to hit Earth in the next two centuries. A goal of this small mission was to understand the structure of the asteroid. Knowing whether it is a rubble pile or a dense single body seems important if you're contemplating blowing it up on its route to Earth. It might make the difference between blowing something to smithereens and merely turning one asteroid on a collision course into an SL9-like chain of impactors, hardly a better prospect.

To answer these questions, OSIRIS-REx was supposed to carefully touch down, making it the first spacecraft to sample a near-Earth asteroid. After several rehearsal feints, on October 20, 2020, it made contact with Bennu. More accurately, it sank its probe into what looks like powdery, loose matter. Images of the probe touching the surface are dramatic, reminiscent of what happens when a small child smacks a sandcastle with a plastic spade. Rocks and dirt flew everywhere, and OSIRIS-REx's sample collection device became jammed with larger pebbles than

The arm of the OSIRIS-REx spacecraft touches the surface of asteroid Bennu. The asteroid's crumbly surface complicated the mission, which aimed to return samples to Earth. *Credit: NASA/Goddard/University of Arizona*

had been anticipated. Stuffing these precious, if unexpected, rocks into its container, the spacecraft headed for home. By the time you read this, the samples collected by this fabulous mission should, just about, be received by labs around the world, allowing us to understand what the largest rocks that might plausibly hit us are like.

This is all scientifically fascinating, but it falls short of the proper test of a mechanism that might save the Earth. Such a test was carried out by the recently launched DART mission, whose target was the double asteroid Didymos and its moon Dimorphos. There have been, in the history of our exploration of space, many subtle and elegant experiments designed to elucidate the history of our Solar System and the Universe within which it exists. By contrast, just occasionally, it also makes sense to simply hit things, and DART was one of those missions. Watching the team who built it celebrate as their small spacecraft, weighing just 500 kilograms, destroyed itself by smacking into Dimorphos at a velocity of 6.5 kilometers per second was deeply surreal. People at those consoles in mission control normally do all they can to keep spacecraft alive, but DART had carried out its task perfectly, hitting within 17 meters of its intended impact point after traveling 11 million kilometers after launch. Before that it had returned images showing a world of dust and rock held loosely together by the asteroid's weak gravity field and ripe for disruption. The similarity to Bennu was striking; it may be that with many near-Earth asteroids we are dealing with flying rubble piles rather than solid objects.

Our lack of understanding of the internal structure of something like Dimorphos is the reason for carrying out the test in the first place. Calculating the effect of a collision between two solid bodies—billiard balls, for example—is easy. When energy can be absorbed by the thing being hit, and material thrown up and away from the target, things get harder. On the one hand, the asteroid might be able to absorb the blow, like a punching bag; but on the other, material being thrown away will create an opposite force, exaggerating the effect on the asteroid. In such circumstances the asteroid acts almost like a rocket, expelling material in one direction to make it move in the equal and opposite one.

The aim of the experiment was to measure the effect of DART's impact by looking for a change in the orbit of Dimorphos. No immediate result was expected to be seen from Earth, but the world's telescopes were watching anyway, hoping for a show. They got one: at the moment of impact, the asteroid brightened considerably, and a dust shell was soon seen moving away from it. A tail of dusty debris, which appeared shortly afterward and eventually stretched for tens of thousands of kilometers, persisted for days, and was visible even in small telescopes.* It was clear that DART had made a difference. Sure enough, it was confirmed a few weeks later that the tiny moon's orbit had been shortened by thirty-two minutes, much more than had been expected.

A promised mission from the ESA, Hera, will arrive later and assess the damage. The success of the DART test suggests that we should be confident we have a plan to deal with potentially

---

* But not, alas, from La Palma in the Canary Islands, where I was sitting helplessly next to a large telescope under the influence of the most easterly tropical storm ever to form in the Atlantic.

dangerous asteroids discovered a decade ahead of impact. So how are we doing with that catalog? We have most of the larger asteroids, those that might cause a dinosaur-scale disaster, safely recorded in the books, and new surveys, as well as citizen scientists like those at Northolt, are working on the smaller ones. There's a real prospect of having most of those that pose a threat safely tracked within a decade or so. Yet our ability to complete this vital task is under threat. Active satellites are part of the problem, but an increase in the number of ongoing missions leads inevitably to an increase in the amount of space junk that litters the most popular orbits. Without an agreement to clean up space, and to insist that satellite operators deorbit disused satellites, looking beyond Earth to the rest of the Solar System and the distant Universe will be increasingly difficult. Our ability to keep our planet safe from incoming asteroids may be seriously compromised by our failure to avoid littering our surroundings.

Chapter 5

PENGUINS OVER VENUS

Venus is Earth's twin. About the same size as our own world and a brilliant, dazzling presence when seen shining brightly against the backdrop of a twilight sky, for many centuries the second planet from the Sun inspired dreams of life, and thoughts that civilizations might be carrying out their business underneath its thick white clouds. Even the largest telescopes on Earth do not show much detail on our neighboring planet's disc, and attempts to interpret the little that can be seen have often been more confusing than enlightening. For example, Venus shows phases, just as the Moon does, a consequence of its movement around the Sun. Over the centuries, observers looking at it as a crescent, when it is at its brightest, saw a strange phenomenon they called the Ashen Light, a glow that seemed to light up the dark parts of the planet. It was believed by some to be from fires chilly

Venusians had made to keep themselves warm during their planet's long nights, which last for many Earthly months. It's true that the life enjoyed by any inhabitants on Venus's surface would undoubtedly be gloomy, with that cloudy blanket blocking any direct sunlight. But it was easy to believe that a second Earth, one warmed by the heat of a sun some forty-two million kilometers closer than our own, must surely be a balmy paradise compared with our often frigid globe. Surely there must be creatures, and campfires, over there, waiting for us to find them.

It was a surprise when the first robotic explorers to visit Venus reported a world that couldn't be more different from that bucolic vision. The planet's atmosphere is laced with sulfuric acid and is a hundred times thicker than Earth's, resulting in both crushing pressure at the surface and temperatures that often exceed 450 degrees centigrade, day or night. The few probes, all Soviet, that have successfully landed on the surface of the planet and transmitted information back to Earth melted after an hour or two. Observations made by longer-lived craft in orbit show a surface devoid of craters, indicating recent resurfacing* by large-scale volcanic activity, and bearing several suspicious hotspots that glow in the infrared and look like active volcanoes. As I'm writing the final version of this text, there's been, just in the last few days, the announcement of a possible observed change in radar images taken a decade or so apart that seems to show a crater filling with lava, confirming the presence of an active volcano. If this turns out to be accurate once new probes arrive at Venus in the next decade, this will be a most

---

* To a geologist, "recently" means "in the last few million years."

significant discovery, not at all minimized by the fact that the researcher in question explained he'd spent time trying to find changes on the Venusian surface during dull Zoom meetings. Whether this particular result holds up or not, I think we can conclude that any travelers unwise enough to select Venus as a holiday spot might have to worry about lava, as well as about being crushed by the pressure of the thick atmosphere, boiled by the high temperatures, and dissolved by the acid while they are asphyxiating.

With the results of the first pioneering spacecraft, Venus became the Solar System's great disappointment, and the idea of any sort of life there quickly became a joke. Though Mars proved on closer inspection to be a barren desert rather than host to a grand canal-building civilization, there remains the possibility of at least simple life under the red planet's surface. On Venus there seemed little hope of any life at all, and when, in the 1970s, Patrick Moore—my predecessor and colleague on the BBC's *Sky at Night*—presented a documentary that slyly poked fun at what he called "independent thinkers" who believed in UFOs, telepathic aliens, and an inhabited, cold Sun, the title he chose was "Can You Speak Venusian?" The belief of one interviewee, an acolyte of the Aetherius Society who claimed he could speak a space language from Pluto, that Jesus came from Venus ranked for Moore as the most absurd of all.*

---

* The program is wonderful viewing. Patrick is nothing other than polite and curious when faced with all sorts of absurd ideas, an interview technique developed years ahead of Louis Theroux. I've watched it many times and was thrilled to meet the current members of the Aetherius Society, still promoting their message of intergalactic harmony when I encountered them at the Royal Festival Hall decades later.

I mention this to underline the incredulity I felt when I answered the phone to Chris North, another old *Sky at Night* comrade and cosmologist and the astronomer in charge of public engagement at Cardiff University. It was a sunny August day, and I was sitting on my back doorstep contemplating the yard and its profusion of life.* Chris is best described as even-keeled, a man capable of keeping almost any situation in appropriate perspective. If a UFO landed in Parliament Square, I'd nominate Chris to go and have a quiet cup of tea with our visitors and sort out any intergalactic misunderstanding, but on this particular day I could hear the excitement in his voice, which cracked as he said: "Well, it's about Venusians . . ."

The story Chris was letting me in on was known at that point to only a handful of researchers. A team led by Cardiff's Jane Greaves had been using radio telescopes, sensitive to short wavelengths, to look for the signature of a chemical known as phosphine in Venus's atmosphere. I've known Jane for years, mostly because we are both enthusiasts for and observers—in my case as a novice and occasional, in hers as a black belt of some distinction and renown—at the James Clerk Maxwell Telescope (JCMT) on the Big Island of Hawai'i.

The JCMT is what astronomers call a submillimeter (sub-mm) telescope, sensitive to the same range of wavelengths used by the microwave in your kitchen to heat water, a little shorter than conventional radio but longer than the infrared. This part of the spectrum is ideally suited for peering into the heart of star-forming nurseries, revealing places that would otherwise be

---

* Mostly weeds. Occasionally a bird.

hidden from view by the dust that surrounds the cocoons of new stars. The JCMT was built by a consortium of British, Canadian, and Dutch funders in the 1980s[*] and has been upgraded since, but by the end of the first decade of the twenty-first century it was struggling for funds. Eventually, operations would be taken over by a consortium of institutions from East Asia, but with funding uncertain before their intervention, there was a need to raise the telescope's profile. Jane was prominent in a band of JCMT enthusiasts who had long looked for novel ways to make use of the veteran telescope—a chat in Uncle Billy's, long the home-away-from-home dive bar HQ for British astronomers on

The structure that surrounds the James Clerk Maxwell Telescope on top of the Big Island of Hawai'i. The dish is hidden by the dust shield, which is—apparently—the world's second largest piece of GORE-TEX, and which is transparent at these wavelengths. *Credit: East Asian Observatory*

---

[*] A period that includes an exciting episode when it was stolen by pirates en route to the Big Island.

the Big Island (now, sadly, closed forever), a decade earlier had led them to search for methane in the fountains of Enceladus, but now she considered Venus.

With more and more planets being discovered around other stars, some of which seem to be promising homes for alien life, or at least possibly harboring the sort of conditions you and I would find acceptable, speculation about how it might be possible to identify life from a distance is increasing in both volume and seriousness. Most of this work is separate from SETI. Talking to aliens would be nice, but when the majority of the thousands of planets we know of are destined to remain nothing more than points of light in even the largest telescopes, the best we can do is to work out what can be inferred about conditions there from a close study of their atmospheres. Identifying oxygen, for example, in the atmosphere of an Earth- (or Venus-) sized planet would be immensely exciting; in our planet's atmosphere, oxygen persists in significant quantities because it is being produced by plant life. Without biological activity, it would quickly become locked up in rocks and thus hidden from view. If we could point at a star in the sky and say that it had a seemingly rocky planet in orbit around it, with substantial oxygen in its atmosphere, a serious prospect for the next decade given the capabilities of telescopes being built right now, we might have identified our nearest neighbors.

It is true that there are other, geological means of producing substantial atmospheric oxygen, at least for a short while. The possibility of volcanic activity, or some exotic atmospheric chemistry, means that if the European Extremely Large

Telescope* finds oxygen in the atmosphere of some nearby rocky world, there will be much argument before the discovery of an alien ecosystem can be universally acknowledged. Oxygen will be a clue, albeit a significant one, that a biosphere might be operating, but other chemical signatures are less ambiguous. Phosphine, for example, a molecule formed by three hydrogen atoms attached to phosphorus, is produced on Earth only by a few obscure factory processes or by life itself, most notably in the stomachs of penguins. This latter property has led to its use in seabird censuses conducted with images from orbiting satellites, as it is possible to estimate the number of penguins present on an ice floe from the strength of the phosphine signature in their liberally scattered guano, but it also makes the compound a powerful biomarker.

Due to its structure, like that of many simple molecules, including water, phosphine happens to interact with frequencies that lie in the sub-mm part of the spectrum, meaning that it can emit and absorb radiation at the frequencies studied by the JCMT. Though Jane and her friends realized that their telescope was unlikely to be sensitive enough to detect phosphine in the atmosphere of a planet around even a nearby star, they felt that at least a proof-of-principle search of the Solar System for the molecule would be fun. It would show that the JCMT was capable of

---

* Originally this was supposed to be the OWL—the Overwhelmingly Large Telescope—before a lack of funding forced a redesign, making it merely Extremely and not Overwhelmingly Large. The ELT is the successor to the Very Large Telescope and also provides yet more proof that astronomers should not name things.

observations at the appropriate frequency, and it would establish the concept of looking for phosphine as a normal astrobiological activity.

Venus isn't an easy target for the JCMT. The fact that Venus is never far from the Sun in the sky caused some problems in the execution of the observations; bouncing sunlight directly off your dish and into your sensitive instruments would tend to play havoc with careful and subtle measurements, and even when that's avoided then the brightness of Venus itself causes problems. Adjustments to software and strategy were made, and the observations eventually went ahead. The team didn't expect to see much in their data, though they hoped to use a null result to say something about the abundance (or otherwise) of phosphine on Venus, at least working out how much phosphine there was not in the scorching, molecule-destroying environment on the planet's surface.

Judging a whole planet by its average temperature can be misleading, however, and is best avoided. Knowing the average conditions on Earth would hardly prepare an alien visitor for being dumped in the Sahara Desert, or in the middle of the Pacific Ocean, or at the South Pole. Travel to somewhere new, and you need to find out what season it is before packing your suitcase, not rely on a yearly average. It turns out that even Venus, hellish on a global scale, may have a temperate zone, high in the atmosphere, where the atmospheric pressure matches that of sea level on Earth, and the temperature would qualify as a balmy summer's day where I'm from.*

---

* Torbay, the self-styled English Riviera. Suggested slogan: Definitely habitable by most life, even in the winter!

Astrobiologists had noticed the potential of this strange environment, high in the Venusian clouds, as far back as the 1960s, long before the JCMT was dreamed of or Jane and her team thought about taking observations. In the early days of the Solar System, it seems possible that Venus may have been much more Earth-like before the heat of the young Sun, the ending of plate tectonics on the planet, or some other process started the runaway greenhouse effect that resulted in its present state. If life evolved on Venus as quickly as it seems to have done on Earth, appearing within the first billion years of the planet's existence, then as the planet's climate changed it is not difficult to imagine life being confined to increasingly cramped ecological niches as things went from bad to worse on the surface. If Venusian single-celled organisms—bacteria or something of that kind—have ever existed, and if they thrive in similar conditions to those of their Earth-based cousins, then the last remnants of a once planet-wide ecology may be clinging on high in the atmosphere, just in this temperate zone. The possibility does seem rather remote, not least because the one Venusian problem that height in the atmosphere cannot cure is the acidic nature of the environment; any putative Venus-dwelling beasties will need to be remarkably acid-resistant, even if the temperature and pressure seem favorable to our kind of life. Finding phosphine seemed a long shot at best.

Having done some of it myself, I can confirm that hunting for molecular signatures with the JCMT can be an unglamorous and difficult business. The radiation received by the telescope is split into a spectrum, a graph that shows the intensity as a function of wavelength. If you stare at something as bright as

147

Venus, much of the light comes from the dazzling planet itself. This is useful if you just wish to confirm that the planet exists (or measure its overall temperature), but Jane's team sought to detect specific molecules, which means looking for the signature of light at wavelengths determined by the structure of the molecule itself. (The structure of any given molecule will determine the number and location of any affected regions; in reality this means much comparison to laboriously compiled lists of wavelengths associated with any given molecule assembled from work here on Earth.)

Spotting a dip in the spectrum at just the right wavelength is a sign that a particular molecule is present and absorbing the background light from the planet. Simple enough in principle, but it is easy to get confused; I spent a month of my time as a PhD student trying to trace the source of an especially confounding dip that appeared in one of my spectra of a star-forming region, only to eventually realize that I'd made a completely independent discovery of oxygen in the Earth's atmosphere, which had gotten in the way. Though the JCMT is placed high on top of Mauna Kea precisely so it can get above most of the water in our atmosphere, which would otherwise swamp its detectors, there is still some interference. At the wavelengths Jane and her team were using, the sky glows brightly and detecting even a bright source against this background is difficult, requiring much careful processing of data.

This meant that a lot of work was needed before Jane and her team could inspect their data. It sat quietly in a forgotten folder for a while, but when they got around to working out what it could tell them, it produced a result that prompted both skepticism and surprise. There was, once carefully processed, a distinct dip visible in the spectrum, so something was absorbing light from the planet, and it appeared at just the right wavelength to be phosphine. The team of people involved in checking and rechecking the data expanded as excitement grew. Emily Drabek-Maunder, then a postdoctoral researcher in Cardiff, remembers being asked if she wanted to look for penguins on Venus, an invitation surely no one with an ounce of joy or whimsy in their scientific soul would pass up. Much work, and much wrangling of the data, convinced the small team that the feature they saw in the spectrum was real, and not some chimera produced by the complex processing they were doing. Unfortunately, they were also far from convinced that what they could see in their data was meaningful. The spectrum from the JCMT was just too noisy to be truly persuasive, the telescope straining at its limits to observe anything at all on a planet it was never designed to look at. More data, from a telescope that was more sensitive at just the right wavelength, was needed.

Such a telescope exists, high on the Chajnantor plateau in northern Chile. Visiting Chajnantor is as close as I've come to being in a truly alien landscape, taking gulps of the thin air high above salt flats that sparkle in the intense sunlight and gazing across an undulating plain surrounded by volcanic peaks whose sides are stained by colored lava flows. When I was there, one of them, Láscar, was emitting a plume of black smoke from its summit cone, looking for all the world like a child's drawing of a menacing volcano come to life. The landscape's volcanically contaminated soils contain so much heavy metal that it is a very bad idea to eat anything grown there, though the ability to mine the lithium that supplies many of the world's batteries is some consolation to the locals. The plateau itself is so high that a special transporter goes to the trouble of bringing any of the multiple dishes of the ALMA* array of radio telescopes, each twelve meters across, down to a more hospitable altitude whenever work needs to be done. Despite the hassle of moving these massive and sensitive pieces of equipment up and down a mountainside, it's a better solution than having a crew attempt complex technical work at altitude; human brains work rather better with sufficient oxygen than without.

ALMA, and its attendant astronomers, is there because the plateau is one of the driest accessible places on Earth, and from here observations in the microwave part of the spectrum can be made with crystal clarity. It took two tries for Jane and her team to win a bid for telescope time to follow up on their Venusian phosphine from this remote spot, but perseverance eventually paid off and they were granted it in the winter of 2018–2019. Their next problem was that, just as at the JCMT, no one had actually tried to look at Venus with ALMA before, and the telescope's systems had to be adjusted to make the observations possible at all; the planet was bright enough at these wavelengths to cause worries about reflected light dazzling ALMA's sensitive instruments. The COVID-19 pandemic also interfered, shutting down operations in Chile before the observations could be completed.

---

* The Atacama Large Millimeter Array, and hence a rare example of astronomers not taking liberties with an acronym. As ALMA is Spanish for "soul," it tends to be pronounced as a word, not a set of letters.

Nonetheless, as it became clear that COVID was likely to cause a long break from observing, Jane's team got to work with the data they did have. ALMA is a spectacularly powerful telescope, positioned on a great site, after all. Maybe there was something in the data that already existed? Jane had recruited the University of Manchester's Anita Richards, who had been responsible for teaching UK astronomers how to use ALMA, and who has more experience than anyone in the country at using the telescope. Late one night, Jane and Anita stared at the plot on the screen in front of them. After the complex dance of data reduction and processing had been completed, ALMA's message seemed as unequivocal as it was surprising. The signal, they were sure, was there. Something really does seem to be producing phosphine high in the clouds of Venus.

The structure of the signal in any spectrum carries data. From the shape of the line, information about the temperature and atmospheric pressure of the environment in which a molecule—in this case, phosphine—exists can be deduced. It turns out that this possible signature of life is found at just the height in Venus's atmosphere where conditions are most like those on the surface of Earth. If life on our sister planet is like that on ours, then the signal matches; we see phosphine just where life would choose to hang out. Rather than being simply a lifeless hellhole, the Venus of the imagination becomes with this discovery a tragic or inspiring place, a once-verdant oasis that would have been populated, presumably, by a variety of species we can only imagine. Single-celled organisms may have multiplied fruitfully in Venusian oceans,* and strange and exotic plants may have carpeted the continents, reaching up to the bright Sun through what would then have been clear, not cloudy, skies. Perhaps in the undergrowth, some larger animal—an alien, but one raised just next door to the Earth—once stirred.

Before our evolutionary imagination can go too far, we have to turn up the thermostat. The result is to thicken and acidify the atmosphere and increase the pressure on the surface. The window of habitability for our kind of life on the surface of Venus may have been as short as two million years, and whether the process that ended this idyll began with planet-wide volcanism, or is merely the consequence of living too close to the Sun for too long, once begun it will have been inexorable and therefore inescapable. Any chance of complex life persisting must have been quickly lost, but, as the discovery of life in unlikely places on Earth has taught us, once established it is nothing but tenacious. It clings on where it can,** with Darwin's great ratchet of natural selection under pressure driving species to change, to solve the problems of living in even the most extreme conditions. High above Venus's scorching surface, algae of the air seem to have survived, living out their strange single-celled existence in droplets, isolated from the acidic environment around them.

Such droplets may even provide the basis of an odd sort of life cycle for our new Venusian friends. As they grow like raindrops, gravity inevitably drags them lower in the atmosphere. Here the

---

* Oceans, incidentally, believed to have been lightly carbonated, making the whole thing seem more like an exotic cocktail than a planet.
** Famously, it finds a way.

temperature may preclude life itself but spores could still survive, just as they do after forest fires and in all sorts of other extreme conditions here on Earth, and they may be able to rejuvenate when the planet's strong winds happen to lift them back up to the temperate zone. A brief life follows for the single-celled organisms they produce, before they are carried once more down to the hellish lower layers of the atmosphere. Sara Seagar of MIT and her group, collaborators of Jane's who are largely responsible for promoting the theory of life in this most unlikely of habitats, have even suggested that strange dark splotches seen in satellite images of Venus's atmosphere taken in ultraviolet light, which seem to come and go over the course of days and weeks, may in fact be these layers of dormant spores. This explanation for the dark patches even predates the phosphine discovery, though it nicely complements the idea of a layer of single-celled organisms living high in the atmosphere and circulating within it.

My *Sky at Night* colleagues and I covered the story of phosphine, of course. How could we not follow up on a tip that might have indicated the first detection of alien life? The result was a very surreal day, wandering what were the still-empty pandemic corridors of Cardiff's Physics Department with Jane, talking about what she had found and what it meant to her, and what such a discovery would signify for our understanding of the Universe.* My biggest worry in heading to Cardiff, head reeling with possibilities, was that the impact of the story would get lost in the technical details, and that Jane's excitement about what they

---

* One of Jane's coauthors, chemist William Bains, as well as pouring acid on a remarkable variety of foodstuffs for the benefit of the camera, brought his guitar and sang us a song about phosphine. It had many, many verses.

might have found, which was so clear in the brief call we'd had prior to filming, would be hidden by her need to be cautious on camera. The paper the team had written was entitled "Phosphine in the Upper Atmosphere of Venus," not "We Found Venusians!" Understandable caution, but much less good television. Watching the film that we made now, you can feel my giddy elation early in our interview as Jane gazes down at her desk in thought, before looking up and saying, clearly, in response to my very first question about what they'd found: "We might have found aliens." To be in on her secret, just for a day or two before the public announcement, was genuinely a privilege. Jane was clear that we didn't understand the chemistry of Venus well enough to even think about jumping to any conclusions, and I finished the show by saying the same thing. We had seen unusual chemistry that might be life, but which might not be. But I also said what the discovery could mean; it was hard, in the circumstances, not to let one's mind run away with the possibilities.

When, a few weeks later, a Zoomed press conference carried news of what the team had found internationally, the discovery was instantly the talk of the astronomical world. Anyone capable of holding a coffee cup had opinions and discussed them with their friends and colleagues. These conversations quickly produced a few follow-up papers. Excitingly, a reanalysis of data taken by the Pioneer Venus spacecraft as it descended through Venusian clouds in 1978, more or less neglected ever since, looked promising. Uniquely among missions that have descended to the scorching planet surface, Pioneer Venus took measurements as it passed through the atmosphere, including some that made use of a mass spectrometer, a device capable of measuring the mass of molecules it encountered. A mass spec can't tell you what molecules it's measuring, just how much they weigh; but there, ignored for over forty years, was a signal consistent with detection of phosphine at just the right height to back up the observations that had been made from the ground.

Other responses were less promising. As discussed above, observations of a source as bright as Venus were difficult for both the JCMT and ALMA, and the data Jane had to hand required a lot of processing to identify the signal that she felt betrayed the presence of phosphine in the atmosphere. Two separate teams, working with the raw data that Jane's team had used, found no significant evidence for the molecule's presence. This seemed bad. For many of my colleagues, it was certainly enough to dismiss the result; as often happens, it appears that a team had made a high-profile discovery, got the publicity that goes along with that, only to have other experts, more cautious and less invested in a particular outcome, show where they went wrong. This is not unusual, just the scientific process carrying out its useful, usual process of self-correction.

Jane stuck to her guns, but things got worse. A little later, the European Southern Observatory—the organization that built and operates ALMA—revealed that the careful attention paid to the processing carried out for these observations of Venus had identified a software problem. There was a glitch in the pipeline, the software that transforms data received by each of ALMA's dishes into data that could be analyzed by researchers, and it had affected the results of Jane's team. It seemed to be curtains for phosphine-producing Venusians; correcting the error, the rival teams once again saw no sign of a signal at the right wavelength.

Certainly not Jane's fault, but an end to briefly held fantasies of Venusian algae and that marvelous microscopic fauna existed in the planet's thick air.

Except for one thing. Jane's team have reanalyzed their data too, using the corrected pipeline, and they still believe that the signature of phosphine is there. The only real resolution to the row will be more data, and as I write new work using data from a different run on the JCMT seems to confirm that some phosphine at least is present. A negative result from NASA's flying observatory, *SOFIA*—an amazing telescope that peered from a hole cut in the back of a jumbo 747 plane, which can fly high enough to escape even more of Earth's water—seemed to scotch our hopes once more, but a reanalysis of that data suggests otherwise. The latest idea is that looking across the whole planet is a mistake, and that the phosphine is best viewed at Venusian dawn, as it seems to dissolve or be disassociated by sunlight during the course of the day. Venusian bacteria, if they exist, may have a brief window of opportunity to live first thing each day, before conditions become too hostile, about as restricted an ecological niche as one can imagine. Still, a day on slowly rotating Venus is 243 Earth days long,* so it's not quite as short a window as one might imagine.

Do these arguments about data and additional detections and nondetections add confidence or make you believe that the whole thing is nonsense? From my office I could throw a pen at people who believe either. We really need new ALMA data,

---

* A day on Venus is actually longer than a year; the planet completes an orbit of the Sun faster than it rotates. Why this should be so is one of the Solar System's mysteries.

but Jane reports that it's proving hard to convince the committee that controls access to the telescope to point it once again at Venus. Until then, we have new JCMT data coming, but it's of course very possible that Venusian phosphine will join the long list of things that turned out not, in fact, to be aliens. It's also possible that in the long run, 2020 will be remembered as the first year that we found proper evidence of life beyond the Earth.

This episode also makes me think that we have perhaps been too restrictive in looking for life only where there is water, and the other ingredients that our kind of life needs. I've already mentioned Cassini's encounters with Enceladus and its water-rich ocean and fountains, but other discoveries with important astrobiological implications also came soon after the probe arrived at Saturn. One of its first tasks was to release the European Huygens probe, leaving this smaller craft free to plummet through the thick atmosphere of Titan, Saturn's largest moon. When previous probes had visited, Titan had appeared in pictures as little more than an orange ball, a thick atmosphere of nitrogen mixed with a complex hydrocarbon haze that prevented us from getting even a glimpse of the surface. This is what makes Titan interesting: it is the only moon in the Solar System with a substantial atmosphere, and the presence of hydrocarbons means that some complex chemistry must be happening despite the low temperatures that prevail this far from the Sun. But it was frustrating not to know what lay underneath.

Cassini's cameras did better than its predecessors, tuning precisely to wavelengths, mostly in the infrared region of the spectrum, where light could penetrate through the cloud and give us a blurry view of Titan's surface. In those initial images indistinct channels and what seemed to be the shore of a lake could be seen, but it was only when Huygens came swinging through the thick atmosphere underneath its parachute that Titan's spectacular landscape came into view.

The first images from the probe revealed Titan to be a world as surely shaped by liquid as our own, with river valleys and lake beds, and large seas and oceans. Unlike on Earth, though, the most important molecule shaping Titan's landscape is methane. Large methane raindrops fall from the moon's cloudy orange skies. Each the size of a lawn tennis ball, they fall slowly enough in the moon's sluggish gravity for them to be easy to swerve if you were out for a walk. Hitting the ground, they form into rivers and feed lakes that come and go with Titan's seasons, evaporating in the summer Sun and then refilling over the course of the moon's long winters. Huygens itself was equipped to touch down either on land or on the sea's surface,* but in the event did so on a frozen plain. This caused some problems for one of its most important instruments, a penetrometer. Not a subtle thing, the penetrometer stuck out from the bottom of the craft in order to prod and measure the resistance of the surface it encountered, hoping to distinguish hard bedrock from a slushy, soil-like surface. The results were surprising. The probe felt an initial resistance before hitting something softer, a sequence characterized by Huygens's principal investigator John Zarnecki at a bleary-eyed press conference the morning after landing as being "rather like crème brûlée." Studying the images sent back

---

* I nearly wrote "on water," carrying Earthly language to a place utterly unsuited to it.

from the landing site, it later emerged that the simplest explanation was that Huygens had hit a hard pebble, one of several that sat on an otherwise smooth surface just where it landed before sinking into the softer layer beneath.* We can even see, in the foreground of one of the images, a large pebble with a crack in it, perhaps the result of a violent encounter with a spacecraft landing on top of it.

Huygens stayed alive for more than an hour on the moon's surface, but it could not talk to Earth directly. Instead, it relied on its mothership to relay messages back, and Cassini was quickly hastening away from Titan. Contact with the probe on the surface was lost when its parent sped out of range almost an hour after the landing, though Cassini did return many times to Titan. On subsequent orbits the scientists and engineers operating its cameras got better at tweaking the instrument settings to look through the clouds and see the surface, and over the course of the probe's seventeen-year mission we were able to watch as lakes filled with methane and ethane as spring turned into summer. Several of these lakes even developed "magic islands" that seemed to appear rapidly before just as quickly disappearing, a phenomenon later attributed to the formation of bubbles as cooling or warming released nitrogen from the mix of chemicals that form Titan's hydrocarbon stew. The presence of these changes indicates the complexity of the chemistry that might

---

* Rumors persist that such was the impact of his crème brûlée comment that the team felt compelled to carry out tests in the laboratory, dropping spare penetrometers into carefully crafted desserts. The results were disappointing; apparently a popadam placed on top of a saucer of curry provided a much better analog to Titan's surface, in texture if not in flavor.

be happening, each heavy rainfall changing the composition of the liquid in the lake. As the temperature on Titan's surface is close to the freezing point of methane, icebergs may also form and melt as the season and the composition of the sea they sit in changes.

This complexity is what makes Titan fascinating to astrobiologists. In such a cold and alien environment, we might not expect anything as sophisticated as Earth's single-celled organisms to be able to form, but that does not mean something close to life might not exist. Life at its most basic is not much more than a set of repetitive and self-perpetuating chemical reactions. Even if nothing as complex as a cell exists on Titan, perhaps the seas do harbor chemistry that fits this definition. Jonathan Lunine, one of the most creative astronomers thinking about other forms of life, has written persuasively of Titan as an edge case, a place where sets of reactions may take place that look rather like the biochemical ones that happen, for example, within each of our cells, and where different sets of reactions might compete for resources and "reproduce," each catalyzing the onset of more. Is this chemical complexity "life" as we would normally define it? I don't know, and I don't think Jonathan does either. The point isn't to engage in semantic debate but to think of Titan as being at one end of a spectrum of complexity, spanning repeating reactions at one extreme to our self-aggrandizing, thoughtful contemplation of the cosmos at the other.*

---

*   Or, of course, we're somewhere in the middle and the all-conquering galactic civilization of Ming the Merciless or some such stands at the other end of evolutionary possibility.

By exploring a different environment, we can expand our conception of what wonders might exist elsewhere in the cosmos. As Steve Squyres, principal investigator of the Spirit and Opportunity rovers, which contributed greatly to our understanding of Mars's past, once told me, we travel to Mars not because it is like Earth, but rather we go precisely because it is different. If you want to study processes that happen on Earth-like planets or look at the possibilities for Earthly life, you're much better off saving yourself the hassle of designing an interplanetary probe and all that messing about with rockets, and staying at home. If you want to be surprised by the possibilities sketched out by the laws of physics and chemistry, let alone biology, you need to go somewhere different, a sentiment surely even more true of the orange skies of Titan than it is of the red deserts of Mars.

It turns out that Titan's lakes and seas, fascinating though they are, cover only a small percentage of the surface. To explore the rest, a little flying probe, Dragonfly, is being built. Dragonfly will spend a few weeks at a time examining a site on Titan before taking off and transferring to another one, conducting a proper survey of this chemically rich but little-understood world. I'm looking forward to finding out about the different parts of Titan, comparing whatever its equivalents of the arid Sahara and the seaside Mediterranean are. Dragonfly is the first mission in my lifetime to explore the surface of a new world;* it will be fascinating to see what it finds.

Recognizing places like Titan as possible homes for life changes the prospects for life in the cosmos. Discoveries made

---

* With all due apologies to comet and asteroid fans.

in the last decade or so have taught us that the Milky Way Galaxy is home to literally billions of planets. Though many—ranging from hot Jupiters that are close enough to their parent stars to be boiling away, to circumbinary worlds with two or more suns in their alien skies—are as far from our own planet as you can imagine, a great deal more attention has been paid to trying to find the most Earth-like environments. In estimating the chances of finding life in the galaxy, working our way through the terms of the Drake Equation, it once seemed natural to consider only rocky entities that are at a suitable distance from their star to ensure a temperate climate, at least from our perspective as inhabitants of a world with a substantial and breathable atmosphere.

There are, admittedly, plenty of those. Recent estimates suggest there are at least a billion worlds scattered throughout the Milky Way that could have a temperate Earth-like climate. That number corresponds to roughly one Earth-like world for every hundred stars. For imaginations newly broadened by thoughts of life in the oceans of Enceladus, or by astrobiological reactions on Titan or by phosphine in the atmosphere of Venus, many more planetary systems become possible homes for life.

Giant planets in other systems may have moons that provide homes for alien races enjoying a spectacular view of their primary; who wouldn't want to relax on a summer's day and watch the slow majesty of a Jupiter-rise? Planets close to their star may have civilizations of the air, not clinging on, like Venus's potential surviving bacteria, but perhaps something grander, with all manner of creatures riding the winds that flow ceaselessly from day to night over such worlds. Planets farther out may have strange communities living under their icy surfaces; would intelligent life that arose in such circumstances have any conception of the broader Universe, or would it remain forever in a cosmos bounded by solid, apparently unbreachable walls? Any story written by science-fiction authors over the years can find a home in this new, expanded cosmos of possibility and surprise.

Carl Sagan would have approved of this new perspective, I'm sure. His much-loved 1980s series *Cosmos* attempted to inspire us to look with fresh eyes at the Universe, combining sequences in which Carl fixes the camera with a gentle yet knowing stare and intones wisdom[*] with less familiar ones involving wandering a spaceship that appears to have all the structural integrity of something found in IKEA's bargain bin. On one of these surreal trips, Carl takes us on a dive into Jupiter, finding hydrogen-breathing jellyfish floating among the ammonia clouds. Is this so much more far-fetched than floating bacteria, high among Venus's clouds?

Science fiction—from *Star Trek*, whose original series included an encounter with the silicon-based Horta, a kind of metal-eating mole, to David Brin's *Sundiver* series with its strange and otherworldly hydrogen empire—has long included life-forms based on a fundamentally different way of life from our own.

Whenever we choose to go somewhere we haven't been before, to look in new ways even at familiar objects, or carry out novel experiments or make new kinds of observations, the

---

[*] If you want to make an apple pie from scratch, you must first invent the Universe.

possibility of discovery opens up. Finding phosphine has reignited interest in understanding Venus's atmosphere, a just reward for an observation that was taken without any real expectation of success, but simply because it could be done. Trusting in chance turns out to be a good idea, even when we turn our telescopes and imagination to the Universe beyond our Solar System and look out at the cosmos.

Chapter 6

STARING INTO SPACE

Long before the first satellites soared into orbit, astronomers dreamed of putting telescopes in space, high above the clouds and with an unimpeded view of the cosmos. It was clear from the first serious reports into the idea, commissioned at the dawn of the space age in the 1950s, that in order to be successful the venture would require a few billion dollars, a lot of patience, and more than a dash of imagination. All of these ingredients have powered the remarkable success of the Hubble Space Telescope, the first ambitious, multipurpose space observatory, into its fourth decade of staring into deep space while orbiting the Earth every ninety minutes or so. Despite its age, the telescope's productivity has never been greater: Hubble still provides us with new views of the Universe, despite relying on a computer less powerful than the cheap laptop I took to college in 1999, while the archive of data that has built up during its long years of ceaseless celestial vigil forms an equally valuable resource—today, more published papers are written from this treasure trove than come from new observations. Now accompanied by the infrared sensitive JWST,* launched on Christmas Day 2021 and providing spectacular, beautiful initial results, hopes are high that Hubble's scientific productivity and ability to astound us will continue for many years to come.

All of which makes it difficult now to remember that a long-delayed Hubble was, initially, a disaster and a laughingstock. Despite planning going back thirty years, the complexity of the technical development necessary to fly a precision instrument in space, and NASA's shifting priorities during a period that saw men walk on the Moon, meant that the telescope was only scheduled for launch in the first half of the 1980s. Even then, further technical problems and the tragic loss of the *Challenger* space shuttle, which blew up shortly after its launch in 1986, killing its seven crew members, caused delays. Hubble only reached orbit in 1990, spectacularly over budget. Though it was successfully deployed from the space shuttle's cargo bay, the first images returned by Hubble's cameras shortly after the astronauts had returned to Earth were not at all what had been expected. The entire point of launching a telescope into space was to get a crystal-clear view of the sky, one unaffected by the

---

* JWST stands for the James Webb Space Telescope, named not for a scientist but for the administrator who ran NASA during the *Apollo* era. The name is controversial because of Webb's presence in positions of power during an era in which NASA discriminated against gay employees; I'll use JWST instead of Webb here. It's a Just Wonderful Space Telescope.

twinkling and turbulence induced by the Earth's atmosphere, which causes our view of the cosmos from down here to be a blurry one. Yet Hubble's first images were no clearer, and in some cases actually worse than those on the ground. The problem afflicted images and data from every one of Hubble's four instruments, and so it was quickly apparent that the main telescope systems were at fault. Astronomy's flagship mission was late, overexpensive—and broken.

Anyone who has looked through the eyepiece of a cheap refracting telescope—the kind sold for a few bucks by chain stores to unsuspecting parents keen to introduce children to the sky, dependent on shoddy optics and inevitably supplied with a wobbly tripod—or even tried to make their own telescope mirror will tell you from firsthand experience that the quality of the optics is everything when it comes to telescope design. Hubble's primary mirror, 2.2 meters in diameter, had therefore been fashioned carefully with unprecedented precision, computer-controlled systems grinding away slowly at the precious disc of glass so that it would be the perfect shape: just right for sending focused light from distant stars and galaxies to each of its cameras. I remember reading reports at the time, before the flaws were found, that told astounded readers that, should the mirror be somehow blown up to the size of the Gulf of Mexico,* the largest bump on its surface would have been no more than a few centimeters high. Despite all this work, it was somehow distorting the light reaching it from the cosmos, producing flawed image after flawed image like a funhouse mirror.

---

* That's 0.64 Mediterraneans for those working in European units.

A repair was clearly needed. Hubble wasn't designed to survive a trip back to Earth, and had been designed from the start to be serviced by astronauts visiting it in the space shuttle. Such missions were supposed to be limited to replacing easily removable components such as the telescope's solar panels, rather than conducting precise optical engineering or diagnosing complex problems. Each shuttle mission cost nearly as much as building a whole new telescope, which meant that it was out of the question to organize a trip to prod Hubble on the off chance that the problem would reveal itself or before a plan to fix it could be devised. Before a rescue mission could be launched, a precise diagnosis was necessary, and so investigators fanned out across NASA's network of contractors and collaborators, determined to check every aspect of the telescope's design and manufacturing.

Remarkably, given the many years of delays between construction and launch, they found the answer to the Hubble trouble waiting for them at PerkinElmer, the Connecticut-based firm responsible for testing the primary mirror, which had to make repeated trips from the place where it was being fashioned to a test rig that was capable of measuring its shape. After each test, changes amounting to no more than a fraction of a micrometer would be made to the mirror's surface, each iteration taking it closer to its final, near-perfect form. Inspecting the test equipment in light of Hubble's poor performance in space, the team found a missing washer, no more than a millimeter or two thick, whose absence meant that the top of the rig was just ever so slightly closer to the mirror than it should have been. As a result, the optical engineers, believing the results of their tests, had with painstaking care spent months dragging the mirror back and forth so that it could be polished by some of the most sophisticated equipment in the world to the wrong shape. And then they had launched it into space, where it now happily orbited far from the reach of those who had made the error that blurred its view of the cosmos.

As its Earth-bound team of scientists and engineers worked on, Hubble continued its scientific observations. However, while much of what the hobbled telescope could do was still useful, especially when it was used to take spectroscopic measurements that were less affected by the mirror's condition than was taking pictures, the lack of spectacular and visually appealing results made it much harder to justify the telescope's cost to the watching press and public. Hubble, and by extension NASA itself, were viewed as failures; I remember seeing a cartoon of a hapless student at a science fair presenting as their project a cardboard box with loose wires hanging out of it. The caption noted that, just like Hubble, the student's project was late and didn't work.[*] Something had to be done.

Now that the cause of the problem had been identified, corrective optics, the equivalent of contact lenses, could be prepared and incorporated in new generations of instruments, but, with no immediate plans to upgrade Hubble's cameras, this wouldn't be an easy fix. Designing, manufacturing, and launching an upgraded set of cameras, which would include bespoke optics designed to correct the primary's distorting funhouse effect, was

---

[*] Hubble, like JWST, was a joint project with ESA and with Canada. It's a source of much angst on my side of the Atlantic that NASA tends to get most of the credit for its achievements; it is certainly true that during this traumatic period it also got most of the blame.

something that would take most of a decade, time that a beleaguered NASA and its international partners could not afford. For an agency whose budget is renewed, and often altered, annually by the arguments of politicians in Washington, DC, the resulting slow drip of failure year after year while it spent a large chunk of its science budget on coping with the problems of its most high-profile telescope was, if not fatal, then hugely damaging.

Somehow the effect of the misshapen mirror had to be corrected before light gathered from distant objects reached Hubble's instruments. That meant putting correcting lenses directly in the path light took to or from the primary, but whatever optics were used to achieve this needed to avoid blocking too much of the telescope's tube; that would cost too much light, hide the faintest objects, and prevent Hubble from seeing the distant Universe it was launched to study. The team quickly realized that it made most sense to let the light hit the main mirror, accept that it would be distorted by it, and then correct it just before it entered each of the instrument windows. More than six separate mirrors or lenses would be needed, each placed in the perfect position to correct one instrument. But how to get them to stay in position?

Wrestling with the problem, one of the engineers responsible, on a business trip in Germany, found inspiration in an unlikely source. Jet-lagged and staring into the hotel mirror bleary-eyed one morning, he reached for the shaving mirror. The small round mirror was mounted on the wall with the traditional, extendable, scissored mechanism that allowed it to be pulled out and positioned just so, ready for a perfect shave. Suddenly, he realized that this modest device was the solution to Hubble's woes:

from a central extendable core, small arms could be deployed, each with a corrective mirror placed in exactly the right place.

The result of this early-morning brainwave was an apparatus called COSTAR, which looks rather like a fistful of dentist's tools, ungainly but effective. It could be incorporated into a unit that slotted in place instead of one of Hubble's cameras, not too hard a task for spacewalkers with their unwieldy gloves. Installed by astronauts visiting the space observatory in 1994 on the space shuttle *Endeavour*, it was a marvel. Almost instantly the telescope's vision was corrected, and where there had once been a blur, clarity appeared. I vividly remember seeing the first images from the repaired telescope in the glossy pages of an astronomy magazine, accompanied by spectacular footage of heroic astronauts risking their lives, seemingly carefree as they casually balanced on the end of the shuttle's robotic arm, which swung them up to and around the telescope.* The whole thing made such an impression on me that I was, for once in my life, almost lost for words a few years ago when, wandering the furthest reaches of the gargantuan Smithsonian Air and Space Museum in Washington, DC, I found COSTAR itself. Not a replica, nor a flight spare, but the real thing, retrieved from Hubble by a subsequent shuttle-servicing mission once the original instruments had all been replaced with newer versions, with their own interior corrective optics. In the cavernous museum hall, its ungainly spread of tiny mirrors looked impossibly fragile, surely too small to be

---

*   The robotic arms on the shuttles, and now on the International Space Station, are the Canadian Space Agency's main contributions to these projects. They provide such a source of national pride that they appear on the country's banknotes. Thanks, Canada!

of any importance; and yet they had been responsible for the iconic images that papered the walls of my childhood bedroom and inspired my love for astronomy and sense of wonder in the Universe.

Think of a Hubble image. (Think of any astronomical image you've seen, and the chances are that it comes from Hubble.) Whether you recall the pillars of creation, the first stellar nursery we saw in detail, the swirl of two colliding galaxies, dancing together across millions of light years, or a crisp image of Jupiter, the odds are good that the image was made possible by these tiny corrective mirrors. I found myself, perhaps a little too enthusiastically, explaining to fellow museum-goers what it was they were walking past, and I recommend anyone visiting DC to go and pay their respects to COSTAR. You'll find it underneath the beautiful golden curtains of the original solar panels from a previous generation of space telescope, which flew as part of the Skylab mission in the 1970s.*

Of all the images produced by the corrected Hubble, one is special. The result of a hundred hours of observation taken either side of Christmas 1995, it changed the course of scientific history, and was born out of the need to demonstrate the unique capabilities of the restored telescope. Yet this most famous image, the Hubble Deep Field, was nearly never taken, having initially failed to capture the imagination of the committee that chooses how Hubble should spend its time. As with most large and expensive telescopes, astronomers around the world have

---

*   For bonus points, find the pigeon trap mentioned later in Chapter 8, which was also on display last time I was there. Tell them I sent you.

to apply for time on Hubble. We write proposals that explain exactly what science would come from allowing us control for just an hour or two.* It is a difficult thing to do, not least because it is often necessary to argue simultaneously that we know precisely what the outcome of a given set of observations will be, so as to reassure those in charge that the precious telescope time will not be wasted, and claim that the science produced will be both groundbreaking and unexpected. So much work goes into these proposals that rejection, often the most likely outcome, hurts. Time is precious, and there are enough good ideas and proposals to keep seven or eight Hubbles and a good number of JWSTs busy.**

Despite the caprices of a system dependent on committees and tired astronomers arguing in conference rooms, it is something of a surprise that the best idea anyone since Galileo has had about how to use a telescope was rejected. Almost since the observatory was first being planned, people had suggested using it to look for ages at absolutely nothing, staring into space for as long as possible in order to try to find the faint light of the most distant galaxies among the darkness. The conditions for such an observation were clear: Hubble's controllers would have to pick a small region of the sky that contains no bright stars, lying far

---

* In most cases, we're not let anywhere near the actual controls. That would be madness.

** The Space Telescope Science Institute, which runs the process, has had in the past to issue gentle reminders that the email address from which such decisions are communicated—blacker@stsci.edu—belongs not to some anodyne server with a silicon heart but to the telescope's technical manager, Brett Blacker, who really doesn't deserve the vitriolic replies some of us have been known to type out and send on receipt of another rejection.

The iconic Hubble Deep Field. Almost everything in this image is a galaxy, not a star. The image is 2.6 arcminutes across—the same size as a lawn tennis ball one hundred meters away. *Credit: R. Williams (STScI), the Hubble Deep Field Team, and NASA/ESA*

from the heavily populated disc of the Milky Way Galaxy, and keep the telescope focused on it for a long time. As the light slowly accumulated in the telescope's digital detectors, fainter objects would gradually appear in the image, almost like a slowly developing photograph.

This "Deep Field" would be our first glimpse not only of the remote Universe, but of the early Universe. Light from distant objects such as those it was hoped Hubble might be able to discern takes billions of years to reach us, so we would view them as they were billions of years ago. The Universe, we now think, is nearly fourteen billion years old. If Hubble could see galaxies whose light had been traveling toward us for ten, eleven, or even twelve billion years, we would be witnessing the cosmos in its infancy, just at the time that the first large galaxies were assembling. For astronomers who had only studied galaxies in their maturity, the idea of getting a look at the first few pages of the Universe's family photo album was thrilling.

If it could be done at all. Many eminent astronomers assumed that, given the size of the telescope and the distance of the targets, any such effort would be doomed to failure, and for good reason. It seemed plausible, and even likely, that typical galaxies of the kind that might grow up to be rather like our own Milky Way would be too faint in their early years to be captured by the observatory. Papers by scientists who were experts in the field argued that Hubble was best used to provide us with detailed views of nearby systems and find new stars in our galaxy; the prospect of it discovering other faint galaxies was remote.

Besides, a hundred hours is a long time. A program that large would keep an astronomer* in business for years, and the risk of coming up with an image that showed absolutely nothing in particular after such an investment, it was decided, simply was not worth it. The bruises caused by dealing with Hubble's misshapen

---

* Or, let's be realistic, given who actually does the work, their graduate students.

mirror were still tender, and pointing the expensive telescope at nothing, only to see nothing, seemed likely to lend itself all too easily to derogatory headlines, fueling a feeding frenzy for politicians with their eyes fixed on NASA's funding, and its budget for doing astrophysics in particular. Attempts to stare at the distant Universe—and Hubble's Deep Field—could wait for the telescope's reputation to recover.

Except for the fact that the program had one fan. The director of the Space Telescope Science Institute (STScI) in Baltimore, which runs Hubble, JWST, and other missions for NASA, thought the plan was interesting, and he was the one person on the planet who didn't have to apply for permission to use Hubble. By long astronomical tradition, a proportion of a telescope's time remains in the personal gift of the incumbent director, as a sort of reward for looking after it. There is a certain logic to this. Sometimes, when dramatic events happen, such as a sudden supernova, the unexpected arrival of an interstellar object, or a comet hitting Jupiter, there is simply no time to apply via the traditional route. In such cases, getting hold of the director and asking for some of their allocation might be the only way to obtain the observations that the community needs, but there is no rule that says that the director can't, instead, use the time for their own scientific interests. Robert Williams was interested in supporting the Deep Field proposal.

A small STScI team worked up a plan. The telescope would observe over the holiday period, and then the resulting image and the data behind it would be revealed to the world almost immediately, at the big meeting of the American Astronomical

Society that is held each January. The deepest image of the Universe ever taken would, hopefully, generate some positive headlines, even if it turned out not to contain very much that was new; and what's more, running such a simple program over Christmas, with Hubble doing nothing but observing the same patch of sky for orbit after orbit, would enable an exhausted team of telescope handlers, operators, and scientists to rest and recover from years of getting their telescope repaired and then working at full speed.

They picked a patch of sky where the telescope could look without worrying too much about the positions of the Moon and Sun, either of which is bright enough to dazzle its delicate detectors. The chosen region, away from the Milky Way and at the time almost completely devoid of interest, was about a twelfth of the size of the full Moon, like a peppercorn held at arm's length. It lay in the northern sky, just below the familiar asterism of the Big Dipper.* Target selected, the telescope and its controllers got to work.

Williams and his team's punt on a deep image of this obscure region paid off handsomely. As a result of these early observations, the patch of northern sky that they selected now has a convincing claim to be the most studied part of the entire celestial sphere, attracting almost unceasing attention ever since the Deep Field was unveiled in Seattle, just ten days or so after the telescope transmitted the final contributing frames down to

---

* For the British, the Plough. On either side of the Atlantic, it really looks like nothing more than a saucepan.

Earth.* Stacked together, the images Hubble took produced a picture of a sky that is hardly dark at all, but speckled with light from almost countless objects. Thousands of them, each of them seen for the first time, were revealed by the telescope's unblinking scrutiny. Three or four stars are visible in the image, easily distinguished from other objects by the spiky pattern, reminiscent of Christmas card stars, imposed on point sources by the design of the telescope's optics; but everything else that can be seen is a distant galaxy, light from which had traveled through the Universe for billions of years before happening to encounter Hubble. Such light must have been streaming past us for all of human history, but now, for the first time, humanity had placed a device in its way capable of capturing and recording it. Through the Deep Field, Hubble acts as a time machine, revealing not the Universe as it is today but the cosmos that once was all those billions of years ago.

Looking closely, one can see almost immediately that this past Universe is different from that of the present day. The Hubble Deep Field's galaxies come in a wide variety of shapes, which go far beyond the neat and tidy spirals and majestic elliptical systems we see today. There are systems lit up by bright clusters of new stars; there are twisted discs, bizarre, sharp-angled systems, and even tiny, gem-like galaxies that seem to be in the process of merging, creating beautiful and sparkling collections

---

* I'm editing this chapter in a bar in Seattle at the end of another American Astronomical Society meeting here, some twenty-seven years after that seminal unveiling. At the end of a review of the first year of JWST by the incredible Jane Rigby, there was a standing ovation that several people likened to the response all those years ago to the Hubble Deep Field.

of jewel-like new stars. Some of the larger galaxies are odd too, with one in particular, which I remember from years of staring at the image, looking like a strange species of porcupine, with spines of star-forming arms sticking up and curving back from a dusty center. Even today, when for me the image is so familiar that I can see most of it when I close my eyes, pulling it up on the screen is an opportunity to get lost in contemplation, marveling at each galaxy's glitter set off against the dark firmament beyond, skipping from system to system as I wonder about the history and present of them all.

Of course, the many careful studies of the images and the treasures they hold tell us much more than what is obvious to my casual tourist's gaze. Galaxies in the early Universe are smaller, less massive, and closer together than they tend to be today, just as we'd expect from a Universe freshly minted in a Big Bang; but they are also, at least in part due to the increased number of close encounters and mergers that their proximity causes, engaged in a frenzy of star formation. You can, in fact, see this in the Hubble Deep Field; it is why galaxy after galaxy is studded with prominent star clusters. This is the Universe at or close to "cosmic noon," a period when star formation far outpaced the activity we see today and everything in the sky sparkled.

These simple observations, made by inspecting a single image, even on their own form pretty good evidence for the Big Bang theory, or at least for the idea that the Universe's story had something like a beginning. Older notions, which often imagine an unchanging cosmos trapped in a steady state, struggle to explain why, when we look at the past, it should appear as different from the present as the Hubble Deep Field does, and it is also true that an early cosmos with galaxies that are closer together than they are now makes natural sense in a Universe that has expanded during the time that their light has been traveling toward us. (The expansion also causes a redshift, a shift in wavelength that we can measure, which is one of the reasons we know that these galaxies are so distant.) These systems, less massive even than relatively modest local, present-day galaxies like the Milky Way, are revealed to be not so much fully grown as, often, the bricks that will end up combining to form larger ones. Indeed, when we look at the colliding systems captured in the Deep Field, we may be seeing that process in progress, the slow assembly of large galaxies under way in its majestic chaos as small systems combine to form larger, more massive ones.

The release of the Hubble Deep Field, which generated enthusiasm from journalists and the public as well as from scientists who were excited to get their hands on the data, went a long way toward the recalibration of Hubble's reputation. It was a beautiful picture that demanded superlatives—the most distant! the earliest!—and which revealed the telescope's unique capabilities and capacity for awe-inducing imagery. It also changed the way that astronomy was done, and taking a deep look into the cosmos is now near the top of the to-do list for any new instrument. The first deep fields from JWST, for example, have already been released, and much of 2022 was spent arguing about which tiny red splotch in the background of an image might really be the most distant galaxy yet seen. Early results seem to suggest that distant galaxies, whose light has been traveling toward us since four hundred million years or so after the Big Bang, are more numerous and perhaps more massive than our standard theories might have expected. Over the next few years, as JWST scans more of the sky, we'll learn whether we have merely not understood these early systems—which perhaps shine with light from stars that are different from those we see around us, or with the glow of material falling onto voraciously growing black holes—or whether their presence and abundance present a scientific puzzle we will have to solve. In either case, they are there, piercing the blackness of the newest JWST images and helping us forget that this telescope, too, was over budget and extremely late in making it into orbit.

In part, the success of the Deep Field came because the fireworks of star formation early on, in these distant systems, were more spectacular a show than had been anticipated, which makes at least some early galaxies easier to observe; but it was also a reward for the STScI team's willingness to roll the dice and try something new. Once it had been shown to work, Hubble soon repeated the trick with a second Deep Field taken in 1998, this time in the southern sky, in the constellation of Tucana. It was a more difficult task for the telescope, as the vagaries of orbital geometry meant it had to deal with occasional light reflected from the Earth, which threatened to wash out some of the fainter sources, but when the images arrived and were processed, they too showed a distant Universe full of small, star-forming galaxies. This proved that what the original Deep Field had found was no fluke: it had dipped twice into the depths of the cosmos and come up both times with a bucket full of galaxies, so we could now be confident that the richness of the original Deep Field was a representative sample of the whole Universe in its first few billion years.

After the Deep Field, and the Deep Field South, came the most ambitious project yet, the Ultra Deep Field, later augmented with new data to create the eXtreme Deep Field, or XDF. Both Ultra and eXtreme* variants depended on the capabilities of new cameras that had been delivered to Hubble by space shuttle–servicing missions, but while earlier astronomers had had to fight to have the telescope take a single deep image, by the mid-2000s it had become integral to the case made to justify building such instruments, especially cameras that were explicitly designed to let us see deeper and thus farther into the Universe's past.

The XDF contains the results of staring for more than two million seconds** at a patch of sky in the constellation of Fornax, which lies just below the familiar form of Orion in the night sky. It shows us galaxies as they were within the first billion years of the Universe's history, the equivalent of me looking back at family photo albums from when I was four, pre–primary school and long before my thoughts turned to space. Not quite far back enough, it's true, to tell the full story of how I ended up the way I am, but there is enough evidence, perhaps, in the bowl haircut and the fact that I'm holding books in many of the images to begin to tell the story of a life. Similarly, though the Universe's dark ages, during which the first stars formed and the galaxies began to assemble, are not visible, what we see is close enough to the beginning of the story for us to draw some conclusions.

Using these deepest of observations, plus data covering the period between then and now, we can see a distant time when

---

*   eXtreme? Ugh.

**   More than 550 hours, but doesn't it sound more impressive in seconds?

galaxies shone brightly, their reservoirs of gas stirred up by interactions with their neighbors to form new stars at unprecedented
rates that have been unequaled since. Such star formation can be
self-perpetuating, in particular when massive stars quickly burn
through their fuel before exploding as supernovae, driving shock
waves into the remaining gas surrounding them and adding to
the glory of these starbursts.

Once, the sky would have sparkled with light from new stars,
each born in enormous and productive stellar nurseries any one
of which would have been large enough to put our famous and
nearby Orion Nebula to shame, but that was billions of years
ago. Since that time, as gas has been used up in most galaxies,
things have gotten quieter.

Knowing the way things are going can turn looking at the
night sky into a melancholic experience if you allow your mind
to project forward into a future in which this trend continues.
If fewer stars are born now than in the early Universe, thinking
too deeply takes us to the point where the last few stars in our
senescent Universe are fading away. But the Deep Fields, I think,
tell a different and more optimistic story. In the same way that
encountering some artifact from an ancient civilization—like
the child's shoe from Byzantine Egypt that sits in Oxford's Ash-
molean Museum, down the road from my office—creates a con-
nection to the past, making it seem very present, so too the kind
of view of the early Universe provided by the Deep Fields renders
that distant time more immediate to me. Rather than comparing what we look at today to the wonders of the past, being able
to see both epochs makes them feel like part of the same glorious
whole, the Universe's story laid out for us to enjoy.

This is all the more true now we've begun to use a different kind of observation to tell the Milky Way's story. Unlike the versatile Hubble Space Telescope, ESA's Gaia satellite and the two small telescopes it has on board have a single job to do. Gaia is a celestial cartographer, the most expert and accurate mapmaking device ever constructed, with a team of scientists who are worthy successors to the medieval astronomers who used all sorts of ingenious instruments to plot the positions of the stars for the first time. Gaia's secret is an exquisitely precise camera, one of the largest ever launched into space, capable of recording the position and, over time, the movements of the nearest billion or so stars. By observing from different points on its orbit around the Sun, Gaia sees nearby stars shift position relative to those that form a more distant background.

This measurement, of parallax, is something you can easily experience right now by holding up one finger at arm's length and viewing it first with one eye and then the other, shutting each in turn so that the finger changes position in relation to the background furniture. Of course, Gaia sees parallax by moving rather than switching between eyes, so a more realistic demonstration would be to have you choose one eye, and then rock your head from side to side to observe how the nearby finger shifts relative to the furnishings of the room that you're in.* The effect is tiny, as the distance between the stars is large; but, once

---

*   I like to get lecture audiences to do this en masse, partly because I think parallax makes more sense if you've experienced it yourself, and partly because the sight of a lecture hall full of people slowly rocking from side to side while staring at their fingers intently never ceases to amuse me.

calculated with care, it provides the most precise measures of distance yet achieved.

Thanks to this effort, Gaia's map of the cosmos is three-dimensional. We can fly around and explore its map of our neighborhood, and, combined with its measurements of stellar movements, we can run the clock backward and forward, watching our surroundings within the Milky Way evolve. Thinking about the future can be enlightening. It is, I think, somewhat satisfying to know that the red dwarf Gliese 710 will make a close approach to the Sun in 1.3 million years' time, passing just a few trillion kilometers away from our star, potentially stirring up the reservoir of comets that live in the outer Solar System and causing a bombardment of icy missiles to pepper the innermost planets. This sort of thing happens every so often, and don't worry: most of the comets will miss us, many will be slung out of the Solar System by Jupiter's gravity, and some of them will look spectacular as they whizz past us. Tell your descendants. Looking backward, astronomers are combing through Gaia's results in the hope that somewhere, hidden in the data, will be the Sun's long-lost siblings, stars born from the same nebula at roughly the same time as our own, long since scattered by the galactic winds.

Larger-scale tides also wash through the galaxy. Several collections of stars, moving together across the disc, seem to be the remnants of smaller galaxies, long since cannibalized by the Milky Way. The exact sequence of events that led to systems known by such baroque monikers as the Kraken or the Gaia-Enceladus Sausage being incorporated into the main body of our disc is not yet clear, but it is certain that the Milky Way has had a violent past. The process of growth by consuming smaller galaxies is one that is still under way, with our current nearest satellites of any substantial size, the Large and Small Magellanic Clouds, trailing behind them streams of stars that have been ripped from them by the gravity of the larger galaxy. A similar stream can be seen behind a smaller system of a few hundred thousand stars, Sagittarius A, for which the process of absorption is almost complete; it has nearly lost its identity completely. We may even owe the existence of our Sun to this particular and peculiar galaxy, our star having formed in what seems to have been a burst of activity five billion years ago, which, we have recently discovered, coincided with a close pass of Sagittarius A to the main Milky Way disc.

Understanding the details of this story will require more time, and more observations of both the distant and nearby Universe. As I've mentioned, our new space telescope, JWST, was designed precisely to follow up on the legacy of the Hubble Deep Field, being launched to fill in the first few pages of the family album. JWST is the most complex uncrewed mission ever launched, and to the great delight of everyone watching on Earth it is working wonderfully. Like Hubble, JWST had its problems before launch, with more than a decade of delays due to engineering problems ranging from malfunctioning camera chips to a rip in the critical sunshield that keeps the telescope and its cameras cool. These difficulties added to the cost of the telescope, the most expensive ever built, and to our nervousness—not least because, in order to host a mirror large enough to capture light from the very first galaxies, JWST needed to be folded up like expensive origami just to fit into the barrel of the Ariane V rocket that took it to space.

JWST orbits with Gaia at a point in space called L2, one and a half million kilometers from Earth, where it can hide our planet, Moon, and Sun behind its sunshield. More than three hundred separate things had to go perfectly for its unfolding to be a success, and, unlike Hubble, it would be beyond the reach of any plausible rescue mission. To the astonishment of those of us watching around the world, and I think even those at mission control, all went well. The first images from the telescope were revealed in June 2022, and among them was an early successor to the Hubble Deep Field. Centered not on a blank patch of sky but on a distant galaxy cluster, it was the product of just eleven hours of observation. Deeper images will undoubtedly follow, but already we can see scattered across the background hundreds and perhaps thousands of newly born galaxies, their light distorted, bent, and sometimes magnified by its passage through the galaxy cluster that sits in the foreground. The cluster's gravity acts as a natural telescope, allowing us to see farther than would otherwise be possible, as does the optimization of JWST and its instruments for the infrared.

It is likely that some of the galaxies captured by JWST are the most distant, and hence the youngest ever seen. It seems certain that this ten-billion-dollar telescope will succeed in capturing the moments when the first galaxies were born, and it has a good shot at revealing the first stars themselves too. If it does so, it will be because more than twenty-five years ago those in control of the Hubble Space Telescope decided to be brave and risk seeing nothing at all.

Chapter 7

LISTENING TO
THE UNIVERSE

In July 1054, a new star appeared in the heavens, shining brightly in the constellation of Taurus, the bull. Positioned not far from the Hyades, the V-shaped cluster of stars that represent the bull's horns, this novelty was bright enough to be obvious to anyone looking at the sky. Chinese and, if our interpretation of the sketchy records that survive is right, Japanese observers recorded its few weeks of prominence.* At the time, such an event would have been mysterious at best and interpreted as some sort of

---

* No mention of the new star survives from Western Europe. Either records have been lost, astronomers here were less attentive than those in the East, or it was cloudy for all of the summer. The supernova would have been visible for a while, making inclement weather an unlikely explanation even in gloomy northern Europe.

portent at worst, but we now know that the appearance of this kind of supernova marks the sudden and violent death of a massive star, a conclusion reached partly because a remnant is still visible. It appears as a misty patch of gas known as the Crab Nebula, thanks to a passing resemblance of the tendrils that appear in deep images to something you might eat, at least in the eyes of hungry, crustacean-craving astronomers.

The Crab's glowing gas envelope, the expanding shock wave produced by the explosion in which it was born, is still expanding nearly a millennium later, scooping up the gas it encounters along the way. Though through a small telescope like the one I have at home it never appears as more than a faint, misty patch, it is a captivating target for amateur astrophotographers, whose images reveal its full shape, as well as for professionals wielding instruments like the Hubble Space Telescope.

Studies of the Crab have helped us understand not only how stars die, but also how such events influence and sculpt their surroundings. Images show how its shells of shocked gas twist and intertwine with each other, their complex forms a legacy of the shaping of the supernova's explosive blast by the progenitor star's magnetic field. But the Crab is much more exciting when, instead of looking at the Universe, we choose to listen, tuning in to the radio waves that reach us from the cosmos. At the heart of the nebula, a steady *dot, dot, dot* beats out, sending a pulse of radio waves in our direction just over thirty times a second, a regular signal from the rapidly spinning, nearly impossibly dense remnant of what's left of the core of the Crab's parent star, which is a few kilometers across. This object, no bigger than a small town but weighing more than the Sun, is all that remains of the giant star that once lived here. This is a pulsar, a lighthouse whose beams of radio emission, powered by the strong magnetic fields that surround the dense core itself—what is known as a neutron star—sweep the cosmos.

The discovery of the Crab pulsar was perhaps the most significant result in the history of radio astronomy, which now spans nearly a century. It was no easy feat: although the Crab is brighter than almost anything else in the sky at these wavelengths, seen with a radio telescope it is still much fainter than a mobile phone transmitting from the lunar surface would be.[*] Attempts to find such faint signals mean that radio telescopes must often be very big indeed. The largest optical telescopes in the world have mirrors roughly ten meters across; a few times a year I find myself standing under the enormous Lovell Telescope, a dish with a roughly seventy-five-meter diameter that dominates Cheshire's Jodrell Bank Observatory, just outside Manchester. (While optical astronomers have to worry about clouds, our radio counterparts can ignore the weather, so northern England becomes a sensible place to put a world-leading observatory.) The enormous dish, the third largest steerable radio telescope in the world, rotates on railway tracks and turns up and down thanks to the rotation of two gun emplacements salvaged from Second World War–era battleships.

The observatory is now one of the UK's few UNESCO World Heritage Sites, and a new visitor center tells its remarkable story, including its years doing dual duty as a telescope and

---

[*] This means that we can say for sure no one is trying to call home from the Sea of Tranquility. Such a transmission would be deafeningly loud for our radio telescopes, and the bill would be enormous.

as an instrument capable of eavesdropping on Soviet activity in the depths of the Cold War. When Sputnik launched in 1957, becoming the first artificial satellite in orbit, Jodrell was the only Western facility capable of detecting the rocket that carried it into space, a modified missile. It has recently been revealed that a skeleton crew from the Royal Air Force (RAF) were stationed on-site throughout most of the 1960s, in case the military needed to take over the telescope. According to the observatory archives, their presence was so secret that, when Russian astronomers came to visit in the early 1960s and they were unwittingly assigned offices right next to the RAF, it provoked much frantic correspondence about soundproofing.

Since then, the telescope has been left to go about its scientific business, undisturbed except for the Bluedot music festival that fills the site for a few days each summer, reducing the Lovell to the world's fanciest projection screen for backing visuals.* It has become familiar to those who live in the area; something this large is easy to spot against the farmland of the rolling Cheshire hills, and from the right angle the Lovell Telescope can be seen from miles around. It is an alien but friendly presence in the landscape, and it seems to have inspired some affection; when it was threatened with closure due to a crisis in science funding in 2007, it wasn't just astronomers who were upset. Schoolchildren wrote protest letters to ministers, local newspapers fulminated, and a charity single to promote the campaign was mooted.

---

* According to Twitter, my dancing was one attendee's "highlight" of last year's festival. I'm not sure this is a good review.

Luckily, all was well, and science still flows from radio waves received by the Lovell's great dish. Though its history is grand, the telescope was saved, ultimately, because the science it is still producing is excellent, not least now that it's connected by fiber-optic cables to six smaller dishes situated across the UK, as far away as Cambridge, 130 miles from the site. Working together allows them to replicate the effect of a much larger telescope, producing sharper images than any one dish could on its own. Joining this network, known as e-MERLIN, has given the Lovell a new lease on life.

Not all of the giant telescopes that marked radio astronomy's first golden age have been so lucky. The enormous Arecibo dish, America's largest telescope, collapsed in 2020, falling into the crater that supported it in the Puerto Rican jungle. Arecibo was a grand thing, a beautiful instrument that used the natural hollow of a depression in the hills to shore up a much larger dish than existed anywhere else, and its loss is mourned by astronomers and the people of Puerto Rico alike. But it did demonstrate that there were plenty of faint objects for ever-larger telescopes to detect, if only we can build systems that are sensitive enough to do so.

The radio telescopes of the future will be built on an even grander scale. The SKA, mentioned in Chapter 1, spans continents and is already under construction. Perhaps even, in the far future, if speculative plans come to fruition, astronomers will build radio telescopes that cover the far side of the Moon, where they can be hidden from the noisy Earth with its humans and their billions of signal-emitting mobile phones. But it is odd to think that all of this activity is due to tinkerers and dreamers who worked outside the formal academy, and who developed the instruments and techniques we use today without any particular scientific goal in mind. Early radio astronomers wanted new telescopes to explore—to look and see what was out there to be seen—rather than building instruments to test existing ideas.

First and foremost among these pioneers was Karl Jansky, an engineer who in 1928 found himself, at the age of twenty-two, set for life with a new job at the expanding Bell Telephone Laboratories. Working in the middle of rural New Jersey at a place called Holmdel, even today not much more than a few prefabricated buildings scattered in the woods, he was assigned the task of investigating the sources of background noise that could interfere with his company's main concern, communication over long distances.

Jansky found that the quiet of Holmdel, far from the radio noise of modern city life, made it an ideal location for his work, and he soon identified the source of the problem. Thunderstorms, it turns out, crackle loudly in the radio. Most of the static at the frequencies that Bell and other companies were trying to use for communication could be traced to the discharge of lightning, which spits radio waves, roiling the ionosphere and producing disturbances that are detectable from a great distance. To show that this was so, Jansky needed to establish not only when interference was happening, but also what direction it was coming from, a measurement that would let him cross-check the arrival of periods of strong radio noise with the known location of active thunderstorms. He therefore mounted his sensitive antennae on a rotating carousel, which, looking more like something from a country fair than a scientific laboratory, quickly gained the nickname of the "merry-go-round." When Jansky spun his carnival ride to point toward storms, it caused the interference signal to gain in strength, just as expected. Most of the mystery was solved, yet a small amount of background static remained that didn't seem to match anything. These faint whispers turned out to be coming from beyond the Earth's atmosphere, and their detection marks the start of radio astronomy. This was a new way of observing the cosmos, born from a desire to improve the practical technology of radio transmission and discovered with an instrument that looked like something made from scrap metal.

Jansky, an iconoclast inventing a new subject from scratch, was operating on his own, and there is frustratingly little about his work that survives in the historical record—no reminiscences from colleagues, or fond biographical notes—to help us understand his personality. I think the measure of the man can be found in the fact that, with this seemingly insignificant final background hiss remaining in the data obtained by his antennae, rather than giving up, assuming perhaps that it could be attributed to distant storms, he kept puzzling away at it, spinning his merry-go-round this way and that, again and again, to try to identify its source. Slowly, as the data piled up, the remnant signal strengthened and faded reliably over the course of a day. As the year wore on, he realized that whatever the cause was it seemed to be moving, tracking the Sun in the sky, following our star on its daily and annual migrations from horizon to horizon.

The idea that the Sun might emit radio waves was exciting and intriguing. Jansky kept monitoring, hoping to confirm that the static picked up by his bespoke merry-go-round was really coming from the Sun itself. Instead, something strange happened. As observations piled up, the source of the radio waves he was seeing slowly shifted away from the Sun's position in the sky. Over the course of a year, the Sun completes a circuit through the zodiac, moving in front of the apparently fixed background of stars as a consequence of the Earth's orbit around the star. As the Sun moved away from whatever Jansky was detecting, what had appeared to be a signal coming from the Sun, it became clear, was much more unexpected, originating from a distant source among the fixed stars.

Further confirmation came when Jansky noticed that the signal peaked in intensity once every twenty-three hours and fifty-six minutes. If you step outside, point at a star in the sky, and stay absolutely still, twenty-three hours and fifty-six minutes later—a period known as the sidereal day—you will find yourself pointing at the same star. We count twenty-four hours in a day because that's the solar day; if you point at the Sun and wait patiently, then it will take twenty-four hours for you to once again be pointing at it. It takes longer because in the time you have been standing there the Earth has moved around its orbit, so the Sun has shifted its position in the sky. This extra shift accounts for the additional four minutes that divide the solar and sidereal days. Though it's only four minutes, this difference between the two explains a lot of navigational and seasonal astronomy. Any given star rises four minutes earlier according to our solar clock each day, a procedure that adds up over time to the slow turning of the seasons and the fact that the winter night sky gives way in turn, four minutes at a time, to those of spring, summer, and autumn.

Returning to the radio sky, the fact that the source Jansky was seeing was fixed in the sky, tracking the stars and not the Sun, meant it was more distant than he had imagined, originating beyond our Solar System. He had found radio waves coming from the Universe.

By observing its position, Jansky worked out he had picked up radiation coming from among the stars of the constellation of Sagittarius, a region that contains the center of the Milky Way Galaxy. You can think of our galaxy as a flat pancake or, as my predecessor on the BBC's *Sky at Night*, Patrick Moore, had it, as two fried eggs "clapped back to back." The Sun lies in the thin disc—the white of the egg—about halfway from the center to the edge. At the galaxy's heart, the density of stars is much greater than it is in our neighborhood, with clusters of young stars shining brightly in a place that still hosts rapid star formation. The whole thing is in motion, and we orbit the center once every 225 million years, a period known as the galactic year. Right in the middle, at our galaxy's dead center, was Jansky's source. It was named Sagittarius A*, or colloquially "Sag-A-star," and we now know that it comes from material in a hot disc around a supermassive black hole at the center of the galaxy, weighing in at four million times the mass of the Sun.

Jansky did not know what the cause of the signal was, and he charmingly called it "star-noise" in the thesis he submitted to earn his master's. Given that this work founded a whole new branch of science, radio astronomy, it must rank as one of the most impressive submissions for such a degree in all of academic history, but it also marked the end of his formal studies.

Understanding the cosmos, or investigating the source of this star-noise emanating from the center of the Milky Way more than twenty-five thousand light years away, was a little too far removed from the commercial business of telecommunications and from Bell Labs' priorities. His bosses turned down Jansky's application for funds to build a much bigger telescope. Meanwhile, the world's professional astronomers—for whom the technical language associated with building antennae and fussing with the electronics necessary to make a success of this new astronomy was utterly foreign, far removed from the mirrors, optics, and observations they were familiar with—were in equal parts distracted, dismissive, and uninterested. They did not know what was causing the emission detected from Sag A*, and by all accounts did not much care. The Great Depression that struck the world's economies in 1929 was no help either when it came to looking for funds to chase this cosmic mystery, nor was the Second World War. As a result, when Jansky died in 1950, his passing was little marked by an astronomical world slow to come to grips with the new view of the Universe he had made possible.

Nearly a century after these first discoveries, astronomers speak Jansky's language fluently; the brightness of objects observed by our modern radio telescopes is measured in a unit named after him. That this happened is in large part due to the work of one of my heroes, Grote Reber, a stubborn iconoclast working, like Jansky before him, not in a fancy, well-funded laboratory or at an august university, but instead building his most important instrument in his spare time on a patch of land next to his mother's house in Wheaton, an otherwise unremarkable part of suburban Illinois.

From these unpromising beginnings, Reber transformed astronomy. While working as a radio engineer in Chicago in the 1930s, he came across Jansky's papers and heard an intriguing, if confusing, broadcast of a recording of the pioneer's mysterious "star-noise" over the NBC radio airwaves. Deciding that this was an interesting area of investigation for someone with his combination of technical skills and boundless curiosity, he wrote immediately to Jansky asking for a job, only to be told that Bell Labs had no interest on following up on the work and were satisfied with the number of engineers distracted by cosmic matters they already had on staff. Undeterred, he wrote next to astronomers across the US, including those up the road at Yerkes, the great observatory in Wisconsin that was then one of the world's leading institutions, only to be met with almost complete indifference.

A lesser man might have left it at that. Reber, however, concluded that the astronomers were "afraid" as they clearly "didn't know anything about radio," and as "nobody was going to do anything, maybe [he] should do something." He lacked collaborators, but, wasting no time, set about "consulting with [himself] and deciding to build a dish."* Soon afterward, working alone,

---

* As with Jansky, there is remarkably little written about Reber, a silence made more inexplicable by the usual American enthusiasm for any project or discovery that started there. These quotes and others in this chapter are from conference proceedings in the 1990s; see the Further Reading section for details.

he had produced a wonder, the world's first true radio telescope: a wooden parabolic dish nearly ten meters across. Once he had it operating at the right frequencies, he could swing it around to point at different parts of the sky, steering it so as to make the first map of the radio sky, which he completed in 1939.

The observations were both difficult and time-consuming, not least because Reber was still commuting to Chicago for his day job. That mattered less than he anticipated, as radio noise generated by, of all things, car ignitions got in the way during the day anyway. Reber would return home after work, sleep for a few hours, and then, once the motorists of Wheaton were safely in bed with their cars parked in their driveways, emerge to observe with his telescope for the rest of the night. Weekends were reserved for data analysis, a painstaking process that led, eventually, to the submission of a paper to the prestigious *Astrophysical Journal* via its editor Otto Struve, who was based at Yerkes.

Reber's results were, to say the least, met with little enthusiasm. His correspondents seemed surprised that such a technical paper could come from someone no astronomer had heard about, working beyond the confines of the astronomical world. Rather than processing Reber's submission in the normal way, Struve and several other delegations of professional men of science decided to pay the Illinois upstart a visit, descending on Wheaton en masse to inspect both the dish and its creator. Despite the habit his mother had developed of hanging her washing on the dish while it sat idle in the day, the visitors found the equipment to be surprisingly and satisfactorily "modern." The submitted paper was eventually accepted by the journal and published, though pared back to the structure of an observational report, shorn of most of Reber's interpretation of his results. Though the grand old men of American astronomy were prepared to grant this engineer the right to publish his data, the other work of science—interpreting what was being seen, understanding it, and gaining the right to be the first to share what it told us about the Universe—was to be retained within the observatories. No mere engineer could be admitted to the inner circle of professional astronomy.

Though the academics probably thought they were doing him a favor by letting him publish at all, their compromise solution understandably impressed Reber not one bit, and he complained that his data had been "sat on . . . until it got mouldy." "Mouldy" or not, the results he wrung from all his effort on late nights and weekends were spectacular. Within a decade of getting the brush-off from both Bell Labs and the astronomers, by sticking to his task he had found the first evidence that the Milky Way Galaxy was a spiral structure. Its long arms were clearly visible in the maps he made of his data, and on top of that he had shown for the first time that, in addition to the diffuse glows that Jansky's results had indicated were present, the sky contained individual, compact sources of radio waves—true radio stars. He had also been the first to detect radio waves from the Sun itself, and in the process demonstrated that periods of increased solar activity could bring with them dramatic bursts of radio waves, the explanation for occasional periods of spectacular interference with electronic systems such as telegraphs that had been recorded for half a century. Until the end of the Second World War, Reber was the only radio astronomer in the world, and he made great use of his leading position to see the Universe as no one else could.

After the war, things changed. The new technology that had been developed for radar systems during the conflict in the UK and the US was redeployed for science, pointing up at the sky instead of monitoring the horizon for approaching aircraft, and other groups began to catch up and overtake Reber. As radio mania swept the astronomical world, he found himself absorbed by the new excitement and was, for the first and only time, actually employed to do research, working on a project to build a new, larger dish commissioned by the US National Bureau of Standards. It was a disaster. Bureaucracy and Reber did not mix well, even when he and the institution employing him shared a goal. He soon became frustrated by the lack of support, and—worse—the fact that no one wanted to put up enough money to build the telescope he planned, which would have been at least ten times the size of his Wheaton effort. Had it been built, it would still be one of the largest telescopes in the world today. The scale of his ambition would not be tamed by the needs or capacity of the government body that happened to be employing him; he felt that a large telescope was needed, and anything else would be a waste of time.

Just as his tolerance with the institution he found himself in ran out, Reber became convinced that the best way to get the telescope to be as sensitive as he wanted was to observe over the ocean. He reckoned that the still water's surface, stretched out in front of his large dish, would help the telescope detect fainter sources in the sky. In pursuit of this goal he went on a trip to Hawai'i, technically a holiday, and then refused to return. That was the end of his gainful employment, and in the end nothing much came of this Pacific adventure; his telescope wasn't built, and in any case the ocean doesn't have the amplifying effect he was hoping for.

Eventually, funded by some modest investments in technology stocks he had made twenty years earlier, Reber moved to Tasmania, where for the rest of his life he continued to work with telescopes of his own design and construction, writing up his results in papers that also featured his own personal cosmology. He didn't believe in the Big Bang, claiming (generously) that astronomers like me follow the cosmological status quo not out of malice or corruption, but simply due to "narrow-minded incompetence,"* and he eagerly awaited the day when data from his next, bigger telescope would show him the distant Universe and prove him right and, once again, the rest of astronomy wrong.

All that stood in the way of his day of personal triumph was the Earth's atmosphere, frustratingly opaque to radio waves at the wavelengths that he wanted to use. Not one to balk at such a problem, or to think small, Reber set about trying to obtain the use of decommissioned intercontinental ballistic missiles.** He wanted them to release liquid hydrogen into the upper atmosphere, creating a window in which his observations could be made. Perhaps fortunately, no one was willing to supply him with an ICBM, but somehow he retained enough clout at the age of seventy-five to get the space shuttle *Challenger* to fire its

---

*   I've been called worse.
**   Obviously.

engines while it passed over his house. He thought it might clear the air, altering the Earth's ionosphere and improving the performance of his latest home-built telescope enough to prove him right. Results, it seems, were inconclusive; however, the shuttle appears to have had some effect on the atmosphere, though not on cosmology.

Reber's later work proved to be more or less an astronomical dead end, but he kept himself busy in a variety of fields, ranging from botany, studying beans in Hawai'i and Australia, to geology, using the genealogy of local indigenous people to determine the age of lava flows. He dabbled in ornithology, rather marvelously finding that Tasmanian parrots are all right-handed,* and built electric cars—in the 1970s—as well as his own eccentrically energy-saving house. The link between these various activities, Reber said toward the end of his life, was that they were "the kind establishment men will have no part of," a disregard for convention that served both him and science well.

None of this activity kept him attached to the mainstream of science that his work had done so much to enable. Long before Reber ended up in Australia, he had been left behind. By the mid-1960s, telescopes in the UK, the US, and Australia had built on his maps, revealing a radio sky sparkling with sources merrily emitting radio waves, though no one yet knew quite what these radio stars were.

In fact, the resolution of the telescopes in operation at the time was very poor. The sharpness of an image produced by a telescope

---

* Right-clawed?

depends on two things: the size of the reflecting surface— the mirror, or in the case of radio telescopes, the dish—and the wavelength of the light being observed. Though radio telescopes like those at Jodrell are much larger than those operating in visible light, they observe wavelengths that are billions of times longer than that of optical light, so their resolution is much worse. We are doomed to see the radio sky only blurrily, and so even as surveys of the sky were being organized to find them, it wasn't even known how large most of the sources being detected were. Opinion was divided between one camp that argued that what the telescopes were picking up were point-like objects, true radio stars, and others who maintained that what was being seen had to be extended objects, like the nebulae and galaxies we pick out when we look into the sky in optical light.

Both options were intriguing, but the idea that something as small as a star could emit radio waves with such power as to be detected from across the galaxy or across the Universe was mind-boggling. Distinguishing between the two options became a question of immense scientific importance in the early days of radio astronomy, and yet it could be tackled using a technique drawn from a nursery rhyme.

As you'll have learned as a kid, little stars twinkle. Actually, stars of all sizes do, their light being distorted by its passage through the Earth's atmosphere on its way to your eye, or to the camera attached to a telescope. The effect is more noticeable when the star in question is close to the horizon, where it cuts a longer path through Earth's atmosphere on its way to you. The effect is much less obvious when the object being observed is not a star but a planet. If you look at a bright planet—Jupiter or Venus perhaps—in the night sky, without optical aid, it will appear as a single point of light. To the naked eye it will be indistinguishable from a bright star, but it is actually a disc a small fraction of a degree across. This steadies its appearance, reducing twinkling. Looking at a planet low in the night sky, I have often been struck by the purity and stability of its light compared with stars nearby; though subtle, the effect is real.

Though it's often not simple in practice, we can, at least in theory, tell planets from stars by seeing if they twinkle, and it turns out that we can use a similar rule to distinguish radio sources that are truly points of light, like stars, from those that are extended, like nebulae. Viewed with a radio telescope, the former should twinkle, though we call it "scintillation" to sound more like serious physicists, and the latter should not. Systematically measuring this effect would, it was reckoned, tell astronomers whether they were dealing with true radio stars, points of light that might represent distant sources, or emission from nearby extended objects.

This important task inspired a bunch of Cambridge astronomers, led by Tony Hewish, to build a new instrument. It was a weird thing, composed of a field full of chicken wire, arranged to create a rough grid. Covering an area equivalent to more than fifty-seven lawn tennis courts, the resulting contraption of poles with wires slung between them looked, as recalled by Jocelyn Bell Burnell, a young Northern Irish PhD student, who built the prototype, "like something you could string peas along." The wires acted as antennae that fed back data to what can only be described as a control shed, which was tucked up against a hedge. It was no one's idea of a normal telescope, and it certainly did not look like a cutting-edge scientific facility, but it was this instrument that was to surprise everyone.

A few bits of the telescope still remain, rusting away slowly on a site that is part of the Lords Bridge observatory, sitting just outside Cambridge on the old Oxford-to-Cambridge railway line. (Other telescopes make use of the leftover railway tracks to move up and down, and the common room is still situated today in the old station building. Visiting today, it isn't hard to imagine the astronomers working with it, poring over reams of paper spat out by the pen recorders that scribble in response to the telescope's signals.) By the winter of 1967, after two years spent constructing the observatory with colleagues, Jocelyn, in charge of analyzing the stream of data, was glad to be able to look at real results. Working alone one cold evening just before Christmas,

The array with which pulsars were discovered, hand strung by Jocelyn and team. It is no longer used; I recently discovered a bit of the array's wiring lives in my boss's office. *Credit: Cavendish Laboratory*

she noticed that the pen of the recorder had, just for a moment, started scribbling frantically, producing what she described as a little patch of "scruff" on the chart.

On its own, this didn't amount to much. The signal was not particularly strong, and as any radio astronomer from Jansky onward would tell you, the sky is alive with noise at these wavelengths. Jocelyn, though, had spent a lot of time in the preceding months with the telescope's charts. At that crucial moment in the freezing observatory hut, she had the feeling that she'd seen something similar somewhere before. Working her way through the miles of archived chart paper, she examined the other occasions on which the same patch of sky had drifted through the telescope's view. The mysterious new signal was not present in each case, but Jocelyn's memory was right: on several previous visits, similar patches of scruff were evident.

Nothing like them had been expected, or had ever been seen before in results from this telescope or any other. A new recorder, capable of preserving more detail, was quickly added to the telescope, and soon its pen was scribbling rapidly as it once again observed Jocelyn's mysterious object. In this new, higher-resolution data the signal remained distinctly odd, consisting of rapid, regular short pulses of radio waves, looking almost like the work of an operator beating out a Morse Code signal and sending it out into the cosmos. The pulses were separated by just forty milliseconds, about the time taken for signals to go from my brain to my fingertips and back as I type this sentence.

Regular repeating patterns are, to say the least, not common in nature, and they were unheard of in astronomy. As the

Cambridge astronomers searched for an explanation, the signal gained the nickname "LGM1," in honor of the Little Green Men who were deemed a possible, unnatural source as they transmitted (repetitive!) messages into space for purposes unknown. In interview after interview in the more than fifty years that have passed since the winter of her discovery, Jocelyn has always patiently insisted that the LGM branding was more of a joke than a hypothesis to be considered seriously, though she does remember cycling back into Cambridge after a long observing session, being annoyed that these putative aliens were getting in the way of her thesis work, which was supposed to be on understanding those distant radio sources.

Joke or not, the Cambridge team realized that they could be on to a major discovery, and even if they did not yet know what it was that they had found, some secrecy was warranted. The level of excitement, and the need for discretion, increased once it was clear that their data held not one single pulsing source but several, meaning a whole new class of object had been found. (The discovery of a second source is remembered by Jocelyn as a relief, as it meant that the phenomenon must be real.) An American observer who visited the Cambridge radio facilities in 1968, before the discovery was made public, remembers noticing on a colleague's office shelf box files labeled "LGM1" and "LGM2," only to find after a moment's distraction that his host had turned them around to obscure the labels.

Before the team could tell the world what they had, they needed to make sure that the signal was real, rather than an artifact caused by some peculiarity of the wiring or the setup of the telescope itself. Such things can confound researchers: the apparent faster-than-light travel of tiny neutrinos from CERN in Geneva to a satellite laboratory in Italy a few years ago, reported to great hullabaloo and skepticism, turned out to be due to a loose cable, which meant the speed measurements were wrong. The one sure test was to try to detect the signals with a completely different telescope, and so one of the other instruments at Lords Bridge was pressed into service. Over the course of a few weeks, its setup was modified to make it sensitive to "LGM"-type signals, and, this done, the team crowded around the recorder as the source passed through the telescope's beam, its field of view.

The difference between discovery and failure, and between success and disappointment, rested on the movement of the pen recorder. Watched carefully by the team as the crucial moment arrived and then passed, it failed to move. No signal was detected. For a few minutes Jocelyn was free to return to plan A, her thesis, constraining the size of distant radio sources, her discovery lost to a historical footnote. Then, five minutes after the predicted time, with the team luckily still there to witness it, the pen moved, tracing the distinctive pattern of Jocelyn's signal. There had been an error in calculating when the signal would transit through the field of view of the telescope, but fortunately it was a small one and the desired confirmation eventually came through. Jocelyn firmly believes that if the calculation had been off by, say, twenty-five minutes, everyone would have gone home convinced that the discovery was nothing more than a problem with the telescope hardware, and her career and astronomical history would both have taken a very different turn.

Instead, the Cambridge team now knew that the signals were real. But what were they? By now, the discovery of several sources had convinced Jocelyn that aliens could be ruled out. The presence of three of the same sort of signal, each coming from different parts of the sky and thus from different parts of the galaxy (or, if they turned out to be farther-flung, from different parts of the Universe), seemed to her to indicate a natural explanation. Some odd astronomical phenomenon should be favored over the possibility that three separate civilizations might have chosen to chatter away in a similar fashion at similar frequencies. I'm not sure this logic holds up, being based as it is on the idea that alien signals need necessarily be rare. If an interstellar communications network was established between distant inhabited planets, even if it was doing something as straightforward as allowing aliens to announce their presence, then surely we should expect some sort of protocol to dictate the form and the frequency of any signal, lest we end up with a cosmic Tower of Babel instead of *Star Trek*–style seamless communication.

Still, needing to invoke the presence of a galaxy-wide communications network to explain what your telescope is detecting does, I agree, at least begin to stretch credibility and seems a little desperate. Luckily, we can be saved from speculation, as before long astronomers had hit on the correct, nonalien explanation for these pulsars, as they were soon called. It was obvious from the start that the rapid pulses of the signal, measured in milliseconds, could only be produced by something small; nothing large could coherently change on those sorts of timescales.

Suitable compact objects were described in the then-nascent theory of stellar evolution. The Crab Nebula, with which I began this chapter, is the result of the death of a massive star, which exploded in a supernova and scattered most of its material throughout the surrounding neighborhood. The core of the star has a different fate. It will, once nuclear fuel is exhausted, collapse to form a compact and nearly unimaginably dense object. For the largest stars a black hole will be the inevitable result, as no physical force will be able to stop the collapse. More commonly, the outcome will be what is called a neutron star, an object so dense that a teaspoon of its material would weigh billions of tons.*

Before its demise, the star would have been rotating. You can use a pinhole projector, or a small telescope with a suitable filter,** to track sunspots across the surface of the Sun and find that the equator of our star rotates once every twenty-seven days. The law of conservation of angular momentum—the same rule that dictates that an ice skater will spin faster if they pull their arms into their sides—means that as the star's core collapses to form the neutron star, its rotation must speed up accordingly. So neutron stars, only hypothetical objects at the time of the discovery of pulsars, were believed to be both sufficiently small, and likely to be spinning rapidly enough, to account for what was known about the LGM signals.

It was quickly realized that such objects must also have strong magnetic fields, inherited like their angular momentum

---

* This is one of those impressive-sounding astronomical facts that astound without actually providing much insight. Can you really imagine what a teaspoon weighing billions of tons would be like? I can't, but as usual with such facts the point is that the number is big. Neutron stars are dense.

** Never look at the Sun with binoculars or a telescope without a proper filter securely in place, as to do so will damage your eyes. This fact was memorably brought home to me early in my career by the scorch marks on the school observatory carpet, caused by carelessly swinging the telescope past the Sun during the day, resulting in a small fire.

from the stars whose death gave them birth. The rapid rotation of these fields can indeed generate powerful radio emission, directed outward along tight beams aligned with the magnetic poles of the system. If the magnetic poles do not align precisely with the planet's spin, then the rotation of the pulsar will sweep these beams across the Universe like a lighthouse, illuminating our radio telescopes again and again and again as they move across Earth. We are like a ship in the cosmic night, our position occasionally revealed by these searchlights, and, like a nineteenth-century ship's crew taking careful note of the distinctive patterns of beams from different lighthouses, we can use them to navigate.*

At least one other radio astronomer had spotted similar signals but didn't persevere in investigating. A decade before, during a public night at the Yerkes Observatory, a visitor, recorded as "a woman from the airline industry," viewing the Crab suggested that its central star was rapidly flickering. Told that all stars flicker, she was insistent enough to be remembered. The sweep of the pulsar beam, thirty times a second, is just about perceivable to a small number of people. Similar stories of pulsar spotting exist from other observatories, but for some reason it is only women who have reported seeing Jocelyn's pulsar directly.

The use of pulsars as landmarks relies on the fact that they are excellent clocks, and from day to day this is true, as pulses arrive one after the other. Many of the world's largest and most

---

* The plaques carried by the Pioneer probes, now heading out into deep space, identify the location of the Earth relative to fifteen pulsars, each labeled with their period. Aliens attracted by the two naked human figures with 1970s haircuts will know where to find us.

impressive radio telescopes are currently engaged in a vast collaborative project to monitor nearby pulsars, using them as a timing array to detect disturbances in space. As we know the period of each pulsar, many of which have been monitored for decades, we can spot if the occasional beat arrives even slightly early or late. If you keep track of enough pulsars, then the hope is that such deviations from the norm can be attributed to the passage through space of what are called gravitational waves, ripples in space predicted by Einstein's theory of relativity.

We have already detected such gravitational waves directly. Any movement of an object with mass will produce a gravitational wave. If you put down this book, stand up, and then jump, for example, you will have disturbed spacetime itself, and a gravitational wave moving at the speed of light will carry the news into the cosmos. Disturbing spacetime seems like a pastime for superheroes rather than for us mere mortals, so it may be reassuring to hear that no matter how athletic you are, or how enthusiastic your jumping is, you will have made only the tiniest of differences to the Universe. Space is stiff; in the technical sense that means you have to give it an enormous whack to get it to wobble appreciably. Detectable gravitational waves are therefore only produced by the most massive objects when they move very fast.

The Laser Interferometer Gravitational-Wave Observatory (LIGO) experiment's successful detection of gravitational waves, which I mentioned in Chapter 1 and which won three of its creators the Nobel Prize in 2017, uses lasers to very carefully look for tiny wobbles in space and has picked up ripples caused by the collision of black holes and neutron stars, violent events that send gravitational waves rippling out across spacetime. LIGO, and its European equivalent, Virgo, have only found waves associated with the merger of objects that weigh a few times the mass of the Sun. These events are fascinating—in one, mentioned earlier, a gamma-ray burst, observed by satellites sensitive to short-wavelength light, was also detected at the same time as the arrival of the gravitational wave. This event seems to have been caused by colliding neutron stars, and studying it has led to a surprising conclusion: in the maelstrom of such cataclysmic events, it seems, most of the gold in the Universe is made. If you have any gold jewelry, then you are wearing the remnants of a cosmic collision.

Despite this success, we have not yet directly seen signals associated with the more dramatic events that must happen when supermassive black holes, each weighing millions of times the mass of the Sun, merge together. Being the product of more massive systems, the wavelengths of the waves that they produce will be much larger than in any experiment we could conceivably build on Earth. This is where the pulsars come in. By monitoring the timing of pulses from pulsars scattered throughout the galaxy's disc, we can turn a good chunk of the Milky Way into a gravitational wave detector.

The trick is to monitor as many pulsars as possible for as long as possible—ideally, for decades. The hope of those conducting such pulsar timing experiments is to detect waves from collisions between black holes that happen after the galaxies that host them collide. Though this involves bodies, each of which may weigh tens of millions of times the mass of the Sun, the signal is hard to find, requiring careful analysis and control over the many confounding factors. To succeed means knowing, for example, exactly how Earth is moving through the Solar System, lest pulses arriving a millisecond earlier or later will tell us only that our position relative to the galactic lighthouses sending out their beams has changed.

Luckily, after much work over the last few centuries, astronomers have a pretty good grasp of how the Earth moves around the Sun, so we can take that into account. Having done so, a few years ago the largest pulsar timing team, a collaboration called NANOGrav, thought they'd found in their data a signature of gravitational waves. Specifically, they thought they saw the shuddering back and forth of the Earth as it is buffeted by the passage of gravitational waves emitted by countless black hole mergers, most of which happened hundreds of millions or even billions of years ago.

Alas, it turned out that this first NANOGrav signal was a false alarm. We didn't know the mass of Jupiter well enough to account for it in the measurements, and the signal the team thought they had found was due to the gravitational pull of this giant planet on the Earth, which has a small but fathomable effect on the Earth's orbit. Luckily, since then the Juno spacecraft has gone into orbit, and its radio messages home have allowed planetary scientists to assess Jupiter's properties in unprecedented detail. Between better knowledge of Jupiter and the pulsar results themselves, the team now knows the location of the Solar System's barycenter*—the center of mass of the

---

* The barycenter is just a little way outside the surface of the Sun. The mass of all the other planets and objects combined is just enough to counter our star's pull that much.

system, around which we orbit—to a precision of about one hundred meters, a shade more than the length of a football field.

A new Jupiter-proof NANOGrav result was released in June 2023, along with similar findings from international collaborations spanning Europe, Africa, Australia, and China. This time things look good, and it seems like the first signatures of merging supermassive black holes are hidden in the data. The pulsars, with their precision timing, provide a fixed grid against which we can see the Earth move; we can sense—using the most modern instruments on the planet and half a century's worth of work to understand the objects Jocelyn found—the Earth's improvised dance in response to the shifting of space as it orbits the Sun. While it may be unsettling to think of our planet crossing a perpetually shifting, roiling spacetime sea, the result is remarkable and provides yet another reason for studying the objects that Jocelyn and the Cambridge astronomers stumbled upon more than fifty years ago.

A couple of decades before the first direct detection of gravitational waves, a special pulsar had already revealed their presence. The discovery of a system consisting of a pulsar trapped in an orbit around another neutron star, known as the binary pulsar,* won the American astronomers Russell Hulse, a PhD student at the time, and Joseph Taylor the 1993 Nobel Prize. The pulsar in this system is such an accurate clock that it can be used to trace its orbit around its unseen companion; as it completes an orbit it moves first toward and then away from us, so the pulses sometimes arrive a little earlier than expected, sometimes a little

---

* PSR B1913+16 not being catchy enough.

later. It was studying these timing measurements that told Hulse and Taylor that something odd was going on, not seen in the behavior of binaries that consist of two normal stars.

The orbital period of the system decreases as the pulsar and neutron star spiral toward each other. The two currently take nearly a minute less to complete an orbit than they did when discovered back in 1974, a decrease that can't be explained by some quirk of orbital mechanics. Instead, something must be removing energy from the system, and it turns out that it's gravitational waves that are doing the job, produced by the motion of the two massive objects. In just a few centuries, enough energy will be lost for the two to crash together and merge, perhaps creating a bright gamma-ray burst for future astronomers here on Earth to detect.

Other pulsars, over the years, have made headlines because of their companions. In 1991, before solid evidence of any planets at all around stars other than the Sun existed, two Manchester astronomers monitoring pulsars with telescopes at Jodrell Bank made a remarkable announcement. Using a technique of looking for changes in timing that was similar to that of Hulse and Taylor, Andrew Lyne and Matthew Bailes had found a regular variation in the steady beat of radio waves that came from one particular pulsar. The effect was small, but it was there in their data, and it seemed to show that this pulsar too was being pulled back and forth by an invisible companion.

Unlike the binary pulsar, though, the companion Lyne and Bailes described was no neutron star. It was a planet, about the same mass as the Earth, the first discovery of a world orbiting a star other than the Sun. With a pulsar hanging above the horizon instead of the Sun, it would have a strange and alien sky and—having presumably survived a dramatic supernova that produced the pulsar in the first place—a very different history from our own; but it was still thrilling to know that another planet existed out there. By apparent coincidence, the length of the planet's year, the time taken to orbit the pulsar, was the same as that here on Earth.

This, it turns out, was not just a lucky chance. In calculating the properties of their new planet, the astronomers had forgotten that the Earth moves around the Sun not in a perfect circle but in an ellipse; and so, as the year of observation rolled on, blips from the pulsar arrived early or late, not because the pulsar was moving but because the Lovell Telescope, Jodrell Bank, and the entire Earth were spinning around the Sun. What was nearly a spectacular discovery was an embarrassing failure to account for the basic properties of our Solar System.

It was an easy mistake to have made, but having announced the discovery to the world with much fanfare, it must have seemed a crushing blow, not least because Lyne as the senior member of the team had to admit their rather embarrassing error. He did so at the biannual meeting of the American Astronomical Society, in front of thousands of his colleagues, and the temptation to spin, to obfuscate, and simply to explain that experimental science is hard—in fact, to say anything other than "I was wrong"—must have been immense; I've watched people in similar situations stand their ground and deny that any problem existed, hide their errors, or argue with their audience. Lyne did not. He just took the simple route and said, "We made a mistake," and as a result two unexpected things happened.

The first was that he got a standing ovation from a crowd who could recognize scientific integrity and honesty when they could see it. The second was a comment from Aleksander Wolszczan, an astronomer who had been following the saga carefully. His team, he announced to the crowd, had been using the same technique on a different pulsar that, it turned out, had not one but two planets. Wolszczan and his colleagues had been prompted to search for planets by Lyne's results, but they had gotten the Earth's orbit right, and so the Manchester mistake had inspired new searches and the discovery of some of the oddest planets in the galaxy. Arguments about how such planets could possibly have survived the supernova, or formed from the debris produced by this event, are still raging today.

Today, more than a dozen pulsars are known to host planetary systems. All are strange, the weirdest of them probably being that which orbits PSR J1719-1438, a rapidly spinning pulsar in the constellation of Ophiuchus. Roughly four times the size of Earth, it weighs more than Jupiter, making it the densest world known. It is believed to be an enormous diamond, the core of what was once a white dwarf star, its outer layers stripped away by the explosion that formed the pulsar it orbits. It is a completely unexpected gem, one of many discoveries that followed from Jocelyn's initial discovery of pulsars, which itself built on the efforts of those like Jansky and Reber who stumbled on a whole new way of finding out things about our cosmos. It is a story that could not take us further from the idea of science as a way of testing known hypotheses; it's one of experimentation, surprise, and delight, enabled by skill and wisdom, for sure, but also by people allowing themselves at each turn to be distracted.

Chapter 8

THE OLDEST LIGHT OF ALL

It is astounding to me that we have observations that can tell us what was happening in the first few hundred thousand years of the Universe's existence. What we can see of the early cosmos tells us that the Universe was then filled with a nearly formless sea of particles and light. Were you somehow to be transported back right to the beginning of things,* you would find yourself enveloped in a glowing fog, making it impossible to see farther than the end of your nose. This strange environment is the immediate aftermath of the Big Bang, the moment 13.8 billion years ago where time got going and our Universe—or, perhaps, this phase of its existence—seems to have sprung into being.

---

* Usual thought experiment caveats about not worrying about how to avoid suffocating in empty space, or exactly what degree of panic kicks in when you are nearly 13.8 billion years older than your own grandmother, apply.

Calling it the "Big Bang" is to apply a confusing label to an astounding idea. The notion that the Universe had a beginning at all was once controversial, and is still counterintuitive. The term "Big Bang" was famously coined by one of the theory's most stubborn and stalwart opponents, Cambridge astronomer Fred Hoyle, who used it dismissively in a radio broadcast. It doesn't sound grand enough to me, and using the expression creates confusion, suggesting, misleadingly, an explosion, something that goes *Bang* within space, rather than the creation of space and time themselves. It's hard when you hear "Big Bang" not to imagine something out of a cartoon, a giant expanding cloud of debris with a pulsating yellow center. What happened at the beginning of our Universe is weirder, and harder to picture, than that, and yet we seem stuck with the term. When, back in the 1990s, *Sky and Telescope* magazine ran a competition asking readers to suggest alternatives, after sifting through thousands of entries the judges decided that nothing better than "Big Bang" had been found. They did commend the use of Calvin and Hobbes's "Horrendous Space Kablooie," but this has yet to catch on.*

Part of the problem is that "Big Bang" tends to be used to refer to two really quite separate things. To the theoretically inclined, and I think to most people, the Big Bang is the single moment when the history of our Universe began, the great Beginning of all things. It's this sense that's being used when we ask questions like "What came before the Big Bang?" (answer, in brief: "Don't

---

* Congratulations to Jacob Jordan of the University of Texas at Austin, whose PhD thesis, entitled "Gooie to Kablooie: The Partial Melting of Planetary Interiors," seems to be the only use of the word in astronomical literature.

know"), "What caused the Big Bang?" ("Don't know"), "Did the Big Bang happen in a particular place?" ("No, all of space was in the same place as the Big Bang, before the Universe started to expand"), or "When was the Big Bang?" ("13.8 billion years ago"). One day we may have a complete theory that explains this first moment and which will tell us whether there have been, or ever will be, other bangs, big or otherwise. It might also go some way to explaining why our particular Universe is the way it is, but at present all we have are some admittedly creative and well-motivated theoretical sketches, suggesting possible routes to it. Such a theory, fully developed, would be the crowning glory of physics, a wondrous tribute to the ingenuity of the scientific mind, but it would also, inevitably, be extremely hard to test.

We cannot run experiments to try to replicate what happened in this most distant epoch. The energies involved in conditions that existed in the Universe just after the Big Bang are immense, far beyond the reach of anything we could dream of matching with a particle accelerator, even if we built one the size of the Solar System. Testing any theory of the early Universe will need to rely on the observations we can make, but we have no way of looking directly back to the beginning. The moment of creation itself is inaccessible to us, trapped behind that luminous fog I mentioned above.

What we can do is make observations that tell us about the very early stages of the Universe's evolution, the first pages of its story. It turns out we can say something sensible about conditions then, and so when observers like me talk about the Big Bang theory, we tend to mean not so much the single moment of beginning but the general and testable idea that, whatever started the thing rolling in the first place, the Universe began its life in a hot, dense state and has been expanding ever since. We can interrogate this idea by looking at what we can see in the Universe around us, and in fact we have a growing body of evidence that it really is how the Universe's story starts. In thinking hard about the Universe's beginnings, we can combine theory and observation to bolster that first sense of a Big Bang—the counterintuitive idea that there actually was a moment when things started.

Start a fraction of a second after the Big Bang itself. This is a physicist's Universe, full of particles, radiation, and energy, governed by simple rules that dictate their interaction, and without the messy complications of things like chemistry or life to get in the way. We do have theories that work well in these conditions, and the equations that describe what's happening in this phase of the Universe's life can be scribbled on a single side of paper. In this sense, understanding the early cosmos is a simpler problem than, say, comprehending exactly how and why a cricket ball spins.* For a complete theory of the cricket ball, you'd have to worry about air resistance, and the effect of moisture, and exactly how the raised seam interacts with a turbulent flow of air over the ball's surface, and a host of other effects. It's a mess. That's OK, as science is often the art of understanding how to approximate a complex situation, deciding which factors actually need considering so that we can make predictions; but the beauty of studying the earlier Universe is that all the relevant

---

* I once explained this on air to Jonathan Agnew, the BBC's chief cricket correspondent, who very much enjoyed being told that spin bowling was more complex than cosmology.

physics can be made immediately tractable and considered at once.

One of the things our theories tell us is that shortly after its creation, the Universe was filled with a sea of electromagnetic radiation. Our cosmos was born in light, and from this radiation we think that both matter and antimatter particles would be spontaneously produced in great quantities, appearing via well-understood quantum mechanical processes as if created from nothing. Having matter and antimatter form together is interesting; as *Star Trek*'s Scotty or any other half-competent starship engineer will tell you, when you combine the two they annihilate, producing light as a result, and so these two processes—the creation and annihilation of matter—are in competition.

As these particles appear, collide, and disappear again, the Universe is expanding around them, with space stretching in all directions. This expansion means that the cosmos is cooling, its energy spread out through the expanding volume of space, and so, soon, as the Universe cools the creation of new matter or antimatter becomes rare. Only the destructive process, which converts matter and antimatter through collision back to light, proceeds with any great speed.

If the Universe had been filled with equal amounts of matter and antimatter, then its history would have been very dull indeed. A brief period of activity would, as all the particles collided, turn into nothing but a sea of radiation. Tidy, perhaps, but a much less interesting cosmos than the one we have, where, for reasons that are still a mystery, it seems that there must have been a slight imbalance baked in from the start. For every billion particles of antimatter produced, our Universe seems to have produced a billion and one particles of matter. Me, you, the galaxy that surrounds us, and everything else you have ever seen are the product of those leftover, one-in-a-billion particles, which have bathed ever since in the light produced by the annihilation of their less fortunate billion sister particles.

The Universe at this early stage consisted of elementary particles, electrons and protons, the building blocks from which atoms are made, but still separate from each other. With particles scattered throughout the Universe, it would be a mistake to talk about empty space at all. Even a few hundred years after the Big Bang, when cosmic expansion was well under way, light could only travel about a centimeter on average before encountering a particle and bouncing off in a new direction. That's why, when I sent you spinning back into this early Universe at the start of the chapter, you were unable to see far, just as the scattering of light off water particles contained in fog makes it hard to perceive even objects that are close at hand. All you see is the surface of last scattering, the fog in front of your face.

Then things changed, rapidly. The most significant single event in the history of the Universe happened four hundred thousand years after its beginning. Astronomers, confusingly, call it recombination. As the universal expansion that had been proceeding since the beginning cooled the Universe, so the electrons within it cooled too. Temperature is a measure of energy, affected by how fast particles move; so this cooling really means that the negatively charged electrons gradually slowed down until, suddenly, they no longer had enough energy to avoid capture by the positively charged protons. The first neutral atoms were formed as the electrons began to orbit the protons; at this stage these new atoms were mostly hydrogen, each containing a single proton and a single electron, but there was also some helium, with two protons and two electrons. The exact ratio of the two atomic species can be predicted from the physical theories that govern the state of the Universe at this time, and it matches what we see in the Universe around us,* a key piece of evidence in favor of the Big Bang idea of a Universe with hot, dense beginnings.

There is another important consequence of the fundamental change to the Universe that happened at recombination. With the electrons tidied away, light could travel freely, right across the Universe. The fog vanished, and in the course of just a year or so the cosmos turned from opaque to transparent for the first time. Light that scattered from electrons an instant before recombination happened is still traveling across the Universe today, and we can detect it, making from it a picture of the infant Universe, frozen at the crucial moment when everything changed, like a crowd on a dancefloor caught in the moment by a camera flash.

The discovery of this ancient light was made by a pair of engineers who were looking for something else entirely. By the 1960s, Bell Laboratories, the company that had so briefly harbored Jansky's early efforts at radio astronomy, was home to a new generation of experiments that explored the potential of long-range communication by radio. Background noise of uncertain origin was, once again, interfering with transmissions. It wasn't always

---

* More precisely, what we see around us is consistent with having started with the correct ratio of hydrogen to helium; stars get their energy by turning hydrogen into heavier elements, primarily helium, and so their work muddies the waters a little, but only by increasing the amount of helium.

there, but when it was the static and hiss made it hard to communicate, and so a new horn-shaped antenna had been commissioned, the pride and joy of Robert Wilson and Arno Penzias, two serious-minded engineers.

A famous photograph of them shows them standing in stiff white-collared shirts and thin dark ties in front of their instrument. It is not beautiful but rather ungainly, fifteen meters across and resembling something a school canteen might provide for the scooping of mashed potato onto plates. At the thin end of the horn, a large wheel is attached to allow it to be rotated, and the whole thing sits on railway tracks so that it can turn to point at any part of the sky. No other instrument looks quite like it. When I visited a decade or so ago, climbing through a fence to reach a site now incongruously located in the rear parking lot of what was then a Nokia research lab,* it felt like I'd stepped back into that picture, with Penzias and Wilson about to emerge from the receiver cabin in black and white.

I was strangely moved by that visit. I was there with two friends and my very new-to-the-world godson, which added to the sense of incongruity as we wandered where we were perhaps not meant to be. I think I reacted so strongly partly because the horn looked precisely like it does in that photo, which I'd seen in book after book; but I was also aware of the instrument's importance in history, its story remembered both because of what was discovered there and by the manner in which it was done. Penzias and Wilson had to work hard at their task

---

* As I'm doing the final edits to the book, there's news that the city of Holmdel might be taking steps to preserve the antenna, making it a more public spectacle. Hooray!

The iconic image of Penzias and Wilson by their Big Horn antenna. The horn fed radio waves from the cosmos to a receiver, which could be analyzed in the hut to the left. The whole structure rotates.

of identifying the sources of interference, scanning the horn around the sky and carefully recording the strength and characteristics of the background noise coming from each direction. It changed over time, and after months of this work and trawling through other data the main culprit was identified. As with Jansky's earlier experiments, the culprits were thunderstorms. The crackle of lightning between the clouds could be easily picked up by Penzias and Wilson's great horn, as well as inadvertently by the equipment of anyone else trying to transmit or receive at the same frequencies.

Identifying the sources of most of this noise was the extent of Penzias and Wilson's mandate. Working out what to do about it, how to filter the effects of thunderstorms or, I suppose, eliminate them entirely, would be someone else's problem. Their careful study of the antenna's data showed, however, that a faint hiss remained once the effects of both nearby and distant weather had been removed. This leftover signal amounted to a total of about 1 percent of the total background noise detected by the receiver, which must surely have seemed too little ever to be a significant threat to even the most subtle of long-distance transatlantic communications. One percent is not much. You or I, scientists and researchers of lesser stripe, would, I am sure, be tempted to call it a day. I can compose the end of the report myself in half a second: "A small remaining source of intermediate noise has not been identified. Further research with a more powerful instrument will be necessary to identify it. However, most of the job has been done . . . ," etc., etc. I've told students countless times to press on with publishing in similar circumstances, prioritizing the need to get work out there over solving that final puzzle. Completing the job can always wait for the next paper. The last mystery tends to be particularly hard, too, another incentive to call the job a good 'un and move on. Luckily, Penzias and Wilson were made of sterner stuff.

At the heart of Washington, DC, is the Mall, the grand parade with the Capitol at one end, Lincoln's magnificent memorial at the other, and the monument to the city's namesake in the middle. Along each side are the grand museums of the Smithsonian Institution, with the Air and Space Museum, dedicated to America's achievements in flight, from the Wright Brothers to the *Apollo* missions and beyond, in a prime position.

In this magnificent temple of flight, if you wander long enough, you might find among the old airframes and astronaut ephemera a strange metal contraption whose place in scientific history is equal to anything that's been to the Moon, constructed by Penzias and Wilson as an attempt to deal with what they thought might be the source of the remaining noise.

They had, according to the description in the paper they later wrote, identified a "white dielectric substance" that was coating the insides of their antenna. The offending material had been deposited, liberally, by pigeons that had made their home within the telescope. The device now in the collections of the Smithsonian is the pigeon trap that was obtained to lure the feathered interlopers out of their scientifically valuable home and, well, let's just assume that they were then released a suitable distance away.* That task done, the antenna was painstakingly cleaned and readied for a final sweep of the sky, which must have been expected, thunderstorms excepted, to give it a clean bill of health. A day or two spent scrubbing a large antenna with toothbrushes should surely have guaranteed the final success of their mission.

After the backbreaking work of cleaning, when the antenna was turned back on, it is easy to imagine its operators slumped in front of the controls as they were denied the satisfaction of an interference-free signal confirming a mystery solved and a job well done. Instead, they were confronted with a continuing and persistent hiss that seemed not to care that the antenna was now clean. The signal was observed to be the same strength whichever

---

* Alas, as pigeons are notorious for returning home, I fear a different solution may have been found.

direction the antenna was pointed in, whatever time of day it was operated, or, indeed, what season it was.

A uniform signal is best explained by a problem with the equipment itself. A nineteenth-century astronomer, for example, who saw giant ant-like creatures when he pointed his telescope at the Moon, realized when he also saw them on the more distant planets, and indeed capering in the vicinity of a few bright stars, that a local explanation—ants in the eyepiece—was more plausible than the discovery of the century. Pigeons, apparently the radio astronomer's ants, had already been dismissed, and no other fault or problem seemed to exist that could explain what the antenna was picking up.

Could it be that this background noise wasn't a problem with the telescope at all? As it was detected wherever the telescope was pointed, it must be coming from all parts of the sky. Any source in the Solar System would stick to the plane in which the planets orbit, following the zodiac rather than covering the entire celestial hemisphere. The Milky Way's disc arcs across the sky, the misty patches visible with the naked eye forming a shape that should be followed by any sources of radiation that live within our galaxy; but what was seen as a background glow detected by the antenna was not confined to this narrow path. The uniform radiation identified by Penzias and Wilson, which seemed to be the same strength wherever in the sky it was found, could therefore not be anything to do with the Solar System or the Milky Way, but instead appeared to belong to the broader Universe, coming from the far distance beyond such insignificant local features. The question of why it should be there at all remained a mystery.

Down the road from the site where Penzias and Wilson were working, at Princeton, a group of cosmologists had the answer. Robert Dicke was a physicist of great breadth and creativity whose work ranged from the fundamentals of atomic physics to persuading NASA to have the *Apollo* astronauts leave a mirror on the lunar surface, so that by bouncing lasers off it our satellite's distance could be determined with precision as part of an attempt to test Einstein's theory of relativity. He was an experimentalist, too, spending the Second World War at MIT, working on radar and other similar techniques, including the first attempt to measure how bright the sky was at microwave wavelengths. We met microwaves in Chapter 5, when they were being used to look for phosphine in Venus. There, the sky shining brightly at these wavelengths was a problem, but for Dicke it was the subject of his wartime research.

His relatively crude experiments did not detect any microwave background; essentially, he managed to show only how dark the sky was to instruments sensitive to such wavelengths, but he returned to the question in the 1960s. By then technology had moved on, and the idea of a Big Bang, or at least of an expanding Universe, was commonly accepted by the scientific community, Hoyle and his collaborators in England apart, and its consequences for our Universe were being pondered by many. The crucial question was what conditions after the Big Bang would have been like. Others had already predicted that if there had indeed been a Big Bang that left the Universe in a hot, dense state, there should be an afterglow of radiation that might still be detectable today. With no sign of this universal background, other possibilities for the Universe's history remained alive. After all, given that we didn't understand the Big Bang itself at all, why should physics dictate that it produce a hot Universe instead of a cool one?

There was also the still-vexed question of why the Big Bang got started in the first place. Dicke had been thinking about the idea of a cyclical Universe, one that goes through periods of expansion and contraction. Each episode might take billions of years to conclude, but each phase of growth would eventually inevitably be followed by a return to collapse. A Big Bang that started each phase of growth would be matched by a later Big Crunch, a compression to match the initial expansion.* According to this model, a presumably infinite set of such cycles stretches backward into the past and forward into the future, creating Universe after Universe without end. I find something slightly reassuring in the idea of the endless dance of the cosmos, and handily it also absolves physicists of the need to explain what caused our particular Big Bang, which merely becomes the latest of countless similar events.**

Dicke was attracted to this neat picture, but he had a problem. By the 1960s, astronomers knew that stars were powered by nuclear fusion. They are, in fact, machines for converting hydrogen into heavier elements, with the Sun creating 596 million tons of helium out of 600 million tons of hydrogen each second.

---

* "Big Crunch" is the standard term for whatever the opposite of a Big Bang is. It's evocative, but nowhere near as fun as the alternative name for this anti–Big Bang: the Gnab Gib.

** Our current understanding of the Universe is that the expansion is actually accelerating, under the influence of a mysterious force we don't understand but which we call "Dark Energy." If it turns out to be the right picture, then this scuppers the idea of a bouncing Universe.

The remaining 4 million tons are turned into energy, which is what causes the Sun to shine. The cores of more massive stars reach hotter temperatures, and so are able to convert helium into still-heavier elements, which are violently shared with their surroundings when they end their lives as supernovae, like the event that formed the Crab Nebula that featured in the last chapter. As a Universe fills with successive generations of stars, it will thus become increasingly polluted with elements other than hydrogen, and a Universe that has existed forever, surviving through endless cycles of Bang and Crunch, must surely by now be running out of hydrogen, the raw fuel for star formation.

Yet our Universe is immensely rich in pure, unsullied hydrogen. Though you wouldn't know it from looking around our own planet—the Earth lacks the gravitational heft that allowed Jupiter, for example, to accrete and then hang on to a thick envelope of light hydrogen gas, thus losing most of its supply early in the life of the Solar System—hydrogen is by far the most common element in the Universe. Something like 90 percent of all the atoms in the cosmos belong to hydrogen, with much of the rest being helium. Where, Dicke wondered, was the debris left over from stars that existed during prior periods of expansion and contraction?

Something about the Big Bang process must be able to reverse the work done by stars in the period between bangs. Dicke and colleagues realized that if the process involved heating the material in the Universe, it would break up atomic nuclei into their constituent elementary particles, no matter what form the material was in just before. A hot Big Bang could act as a cleansing fire, removing the history of whatever came previously

235

and allowing the Universe to start afresh each time with a new supply of nearly pristine hydrogen. Hydrogen, after all, is just protons and electrons. (Conditions in the split second after the Big Bang were, we now think, much more extreme than Dicke and others considered, but the basic point stands: a hot Big Bang is an eraser of history.)

Working with cosmologist Jim Peebles among others, Dicke started to think seriously through the consequences of a hot Big Bang. In particular, they considered what conditions would be like in the particle soup that existed just after such an event, describing in detail for the first time the situation with which I began this chapter. In such a Universe, full of fast-moving particles, light could not travel far without bouncing off a nearby electron, but after recombination, as we have seen, everything changed. Light scattered just before the electrons were captured to form neutral atoms would suddenly have found itself traveling freely across the Universe, and it will—unless staggeringly unlucky enough to have collided with a planet or grain of dust on the way—still be doing so today.

Looking out at the cosmos, we should therefore see this light coming from all directions. Dicke and company calculated that, due to the expansion of the Universe in the billions of years since the Big Bang (to them, the most recent Big Bang), such light would now be most easily detected in the microwave spectrum, if only a receiver sensitive enough could be built. They were just beginning to campaign for funding for a dedicated experiment to detect this expected microwave background when news reached them of Penzias and Wilson's discovery. The Holmdel horn was powerful enough to pick up the faint hiss of the early Universe, even though it worked at wavelengths that were slightly less than ideal for the task. The story is that Dicke, hearing about a faint mysterious microwave signal coming uniformly from all parts of the sky, put down the phone in the lab and announced: "Boys, we've been scooped."

The paper written by the Princeton group explaining what had been found did at least appear in the journal next to the short note from Penzias and Wilson announcing their discovery. The two engineers later won a deserved Nobel Prize for their accidental discovery of the oldest light in the Universe. Though their work convinced most astronomers that the hot Big Bang model was accurate, no matter how counterintuitive the idea of a Universe with a fiery beginning might be, Wilson later explained they'd only realized how big a deal their discovery was when they read about it in the *New York Times*.

Since these pioneering days, studies of what we now call the cosmic microwave background radiation, or CMB, have taught us more about our Universe's story than any other set of experiments. By observing it carefully we can make a map of what the cosmos was like just four hundred thousand years after the Big Bang, and use that to understand how we got from there to the complex cosmos, the network of galaxies and galaxy clusters that fill the sky around us today. In particular, the CMB has proved extremely useful in enabling us to compare the glimpse of the early cosmos with what we see around us in the present-day Universe, measuring the expansion of the Universe that has happened while the light has been traveling toward us.

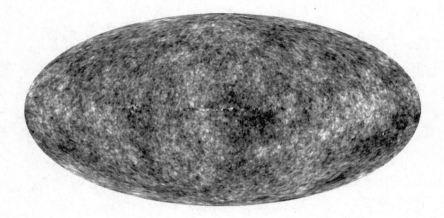

A modern image of the cosmic microwave background from NASA's WMAP probe. Dark and bright spots, representing regions of different temperatures and density, cover the whole sky. Our galaxy's influence has been removed. *Credit: NASA/WMAP*

For example, an obvious feature of the CMB is its uniformity. One of the clues that the antenna was detecting something in the very distant Universe was, as I said earlier, the fact that Penzias and Wilson found its signal had the same strength wherever they pointed their instrument. As the CMB was studied further, it proved very difficult to find any variation in its strength or properties at all, which is confusing. When we look out into today's Universe, we see that the distribution of matter is quite different, and far from smooth. Take a decent-sized telescope on a spring evening in the Northern Hemisphere, a time when our view is directed out of the disc of the Milky Way, and point it at the zodiacal constellation of Virgo or at neighboring Coma Berenices. If the sky is dark enough and the telescope sufficiently powerful, your view will be filled with the sight of faint fuzzy blobs. These are not stars. Photography, which shows more detail than the eye can see, reveals them to be elliptical galaxies, massive systems of stars that populate the heart of the two near-est clusters to us; the Virgo cluster alone contains several hun-dred galaxies larger than the Milky Way, and its gravity exerts an inexorable pull on our own system and its neighbors.

The Virgo cluster is, in turn, being attracted by a still-larger conurbation of galaxies, which lies frustratingly behind the dens-est part of the Milky Way as seen from Earth, making it hard to map. It goes by the wonderful name of the "Great Attractor," not a nineteenth-century circus act but our nearest supercluster. As we have mapped more of the cosmos, we've found that it is organized around such objects, connected by filaments of gal-axies that separate vast regions, or voids, devoid of pretty much anything. Press releases usually talk of our existence within a vast "cosmic web," but I prefer to say that the Universe is sim-ply lumpy. Alternatively, you can think of the cosmos as having grand cities, such as Virgo, and the occasional megalopolis, plus its share of rural areas with relatively little population.*

Studying the CMB asks us a fundamental question: How do you produce a lumpy Universe, filled with environments of such variety, when you start from the smooth cosmos that seems to be revealed by observations of the cosmic microwave back-ground? It was quickly recognized that this was a problem, and it was perhaps the last major objection that stood in the face of the mounting evidence in support of the Big Bang model. More careful mapping of the CMB was clearly needed, and for that it made sense to send instruments into space, where they could

---

* In this picture, we in the Milky Way are firmly suburban, safe in commuter territory like some sort of cosmic Croydon (a notoriously unloved suburb of London).

observe this ancient light free of interference from the Earth's atmosphere.

The first CMB-mapping satellite was NASA's COBE,* launched in November 1989. Its first map of the sky, released in 1992, was a revelation. The previously uniform surface of the cosmic microwave background was shown for the first time to contain tiny imperfections, which showed up in the form of regions with slightly different temperatures. The changes seen between one patch of sky and another were small, amounting to a difference of just one part in ten thousand, but the blotchy map that COBE produced was significant. These small differences in temperature correspond to very small differences in density; slightly colder regions contain slightly more stuff, and thus slightly more matter, than their warmer surroundings.

These small variations turn out to be the seeds for the clusters and superclusters we see around us today. Over the course of the more than thirteen billion years since the light we detect today as the CMB set off on its journey, gravity has been working on these tiny fluctuations, exaggerating their differences. Imagine being a particle in this early, nearly but not quite smooth Universe. You are subject to the gravitational pull of your surroundings, but on average those regions that already have more matter than others will have a stronger pull on you and you will fall toward them, adding your mass to theirs and increasing further the difference in density between neighboring parts of space. As in life, in this gravity-dominated cosmos the rich get richer and the poor most often poorer. In a process

---

* The COsmic Background Explorer. Not one of NASA's finest acronyms.

that we can now watch through beautiful, detailed supercomputer simulations, what starts off looking to the eye like a nearly perfectly uniform sea of matter changes over time. As gravity works, the cosmic web appears from its humble beginnings, able to shape the distribution of the galaxies we see today.

The details of this process tell us a lot about the Universe. As we can observe how things start off in the fluctuations first detected by COBE in the CMB, and can measure the lumpiness of the present-day Universe, we can work out how effective gravity has been during all the time that has passed between these two epochs. This is just a matter of matter: the more stuff there is, the more important a role gravity has to play. The more important gravity is, the greater the exaggeration in density it produces, and the more stark the difference will be between the cosmic web's haves and have-nots, its most and least dense regions. A Universe with more matter in it will end up lumpier than one with less.

Supercomputers have made it possible to run a version of this experiment without the fuss of creating actual new universes.[*] Some compromises do have to be made, as keeping track of the behavior of each atom in our observable Universe is beyond the ken of any conceivable computer; but if we zoom out a bit and ignore complications like chemistry and much of the detailed physics that happens within galaxies, then the task of studying gravity on large scales isn't too bad. My laptop can do an OK job, as long as I don't ask it to do anything else for a week or two, starting with a distribution of particles inspired by the patterns

---

[*]   Which is expensive, takes ages, and is hell to clean up after.

seen in the CMB and keeping track of their mutual attraction over billions of years.

If we make such a simulation of a Universe with less matter sprinkled into the recipe, it'll give a much smoother result, with the differences between cosmic voids and clusters a great deal less pronounced. Add more matter and the opposite happens, with gravity more important, and more galaxies crammed into larger and larger clusters. Somewhere in between, if we get it just right, we can produce something that looks like our own home. If what we're doing in the simulations is writing a recipe for creating our Universe, then following it is more like baking—where precise measurements and careful attention to quantity are necessary—than the kind of slapdash student cooking I'm used to.*

It turns out that the recipe for baking a cosmos to match our own perfectly is a surprising one, needing more matter than might have been expected from looking around the local Universe. If we were to record in a modern version of the Doomsday Book all the stars, galaxies, and planets; all the black holes, neutron stars, and white dwarfs; and all the electrons, protons, and atoms; and then run the simulation with them all accounted for, then we would find that the result is a Universe that is far too smooth to be a good match for our own. Either gravity doesn't work the way we think it does, or there must therefore be more matter in the Universe than we expect.

This discovery is one of the pieces of evidence that have led us to believe that the Universe is filled not with ordinary matter

---

* Add more garlic? Sure, why not? And I'm sure that cream in the back of the fridge is still fine.

but with a mysterious substance, to which astronomers have given the name "dark matter." These particles, whatever they are, must account for six-sevenths of the total mass in the Universe, and though they make their presence felt via gravity, they must not interact with light, or easily with ordinary matter. This creates unsettling thoughts. For example, hold your hand out in front of you. Our best theory of the Universe suggests that tens of millions of mysterious dark-matter particles must be passing through it each second, without leaving a trace or affecting any of the atoms in your body.

All of them will, most likely, pass through the Earth and out the other side without troubling an atom. Just occasionally, if a dark-matter particle scores a direct hit on an atomic nucleus, the resulting rebound might, in the right conditions, be detectable, a prediction that has led an international team of scientists to install successively larger tanks of the inert gas xenon in the underground Gran Sasso laboratory in Italy, which was mentioned in the Introduction. The latest of these experiments, XENONnT, has a tank more than a meter across and a meter and a half high, containing 5.9 metric tons of the noble gas, cooled to its liquid form. Over the course of the five years that this latest experiment will run, the team are hoping that maybe a handful of real events will be detected, producing the first direct evidence that dark matter, which we need to explain the Universe's appearance, actually exists.

Should the XENONnT experiment fail, it won't rule out the idea that the Universe is mostly made of dark matter. Though such a result would obviously be disappointing, it is perfectly possible to imagine a particle capable of shaping the Universe through gravity but that interacts with normal matter so little that no similar experiment could ever find it. What we see in the sky suggests that dark matter exists, but it offers no guarantee that we will be able to identify it with experiments here on Earth.

This makes for a perhaps frustrating situation, but the alternative, tinkering with our theories of how gravity works, is fun but difficult to get right, not least because we have evidence from observations on other scales as to how dark matter affects things. Spiral galaxies like the Milky Way, for example, are rotating so fast that, were they to be made only of the visible matter that we can detect, they would fall apart, a discovery due in large part to the great American astronomer Vera Rubin. Theories that adjust the way that gravity works to account for the motion of galaxies without needing dark matter tend not to produce a convincing explanation for the comparison of the CMB and the present-day large-scale structure that we see around us; and if you start off by fiddling with Einstein's work to explain the CMB results, you often find that you've managed to render, say, such desirable features of the Universe as a stable Solar System impossible. We seem stuck with some form of as-yet-unknown dark matter.

Yet dark matter is, perhaps, the lesser of the problems suggested by the CMB. When thinking about the early Universe, even simple observations can be profound. The uniformity of the CMB has another, more fundamental, message for us. Even as COBE and successor satellite observatories like NASA's WMAP and the European Planck satellite have mapped the fluctuations of the CMB at more frequencies and in ever-increasing detail, it remains true that the view of the sky at these wavelengths is, on larger scales, uniform. What is meant by "uniform" is a little technical, but essentially the idea is that you could cut the map of the CMB in half along any axis you like—down the celestial equator, following the line of Orion's belt, from north to south, however you fancy—swap the two halves over, and the resulting map will be statistically indistinguishable from the real one. The pattern of fluctuations we see will have changed, but its properties will not have.

Play the same game, but cut the map into quarters, eighths, or sixteenths, swap them over, and you'll get the same result. On the scales probed by our observations of the CMB, each part of the Universe is just the same as any other. This may not seem a problem worth grappling with over your cornflakes, if you have until now never let the problem of this isotropy, or uniformness, of the cosmos bother you. But observed facts require explanations, and there seems to be no especially good reason why whatever still-obscure physics governs the Big Bang and its outcomes would have had to produce a Universe like this.

We do, in fact, think we have an answer. The Adler Planetarium in Chicago is a marvelous place, with treasures that include Edwin Hubble's championship-winning basketball,* the *Gemini 12* space capsule, and, carefully sealed and lit with the reverence normally due to a medieval illuminated manuscript, a lined notebook of yellow paper, covered in neat jottings and equations written by a steady hand in blue ballpoint pen.

---

* As it was taken, for fun, along on a mission to service the Hubble Space Telescope by astronaut John Grunsfeld, a fellow alumnus of the University of Chicago, I think this is the only basketball to have been to space and back.

This is the notebook used by theoretical cosmologist Alan Guth in December 1979, and it's open at a page headed by a box surrounding the words "SPECTACULAR REALIZATION." In it, Guth describes the idea that allowing a very short period of extremely rapid expansion, just in the first tiny fraction of a second after the Big Bang, could solve several problems in our understanding of the Universe. Among them is the apparent uniformity seen in the CMB. The cosmic inflation Guth suggested, which is now a standard part of most cosmologists' idea of how the Big Bang works, would have smoothed out any initial differences left after it happened, turning what could have been a lumpy and varied landscape into what we see reflected in the CMB.

Modern cosmology is searching, through new CMB telescopes at the South Pole, in the Atacama Desert, and elsewhere, for evidence that this inflation—Guth's rapid expansion—really did happen. As yet, it remains a good idea to wait for observational proof. Unlike the other chapters in the book, which contain stories of observational surprise and accident, Penzias and Wilson's accidental discovery has led us to the realm of traditional physics. As with the early Universe, things are simple enough for us to write down equations and predict behavior, working as conventional physicists do. In the idea of inflation, for example, we have a hypothesis, and much effort and not a small amount of money are being invested in testing it, with large teams around the globe working together* to create new

---

* There are so many teleconferences and Zoom calls. Seriously, I think this sort of science is mostly teleconferences.

knowledge. It will be a glorious day if they succeed; evidence for inflation would be one of the great triumphs of twenty-first-century physics, and a clue that we are on the right track with our idea of a Big Bang—a remarkable example of the scientific results that can be achieved by following through, relentlessly, on an idea. But all of this activity started because Penzias and Wilson, with their thunderstorm-afflicted antenna and their pigeon traps, stumbled on the oldest light in the Universe and decided to be surprised by it.

Chapter 9

THE CHANGING SKY

I'm a big fan of office doors. Not so much the doors themselves, it's true, but the accumulation of cruft that seems to grow on them over time as favorite pictures, headlines, and cartoons get added to a display. I think it's a scientist thing. When members of staff in institutions like mine keep the same offices for decades at a time, the yellowing paper and out-of-date news stuck with Sticky Tack to the door often tells a story about a career well lived. One of my colleagues in Oxford has displayed the same dismaying graph about reducing funding for graduate students for years now. Other people's doors bear beautiful spectra they took years ago, or an image of the center of the galaxy they're especially proud of making. The same cartoon from the often quirky and occasionally profound webcomic XKCD, which uses a logarithmic scale to show a map of the entire Universe from cosmic microwave background down to the surface of the Earth on the same plot, appears in our building in three separate locations. A mild-mannered and modest senior colleague, who would never think of mentioning her accomplishments in person, has had a picture of herself with President Obama on her door for the last decade or so.[*]

I'm not the only one who pays attention to doors. When the Science Museum in London, a few years back, tried to make an exhibition commemorating the discovery at CERN of the Higgs Boson, they had a problem. The Higgs may be the particle whose discovery completed the Standard Model of particle physics and which is responsible for helping give everything mass—and its identification the particle physics breakthrough of a lifetime—but it is otherwise distinguished by not being visible and being highly unstable on its own, and for both of these reasons, as well as its tiny size, it is hard to display in an exhibition.

The exhibition designers solved the problem not by displaying a particle on a pedestal but by creating a replica of one of CERN's corridors, complete with doors and, carefully reproduced, the associated junk. To be fair, its offices have rich pickings; the place generally looks like it hasn't been touched since the 1970s. A friend who works there now told me of frequent conversations with visitors, especially media crews, who assume the world's premier particle physics lab must be mid-renovation and have to be gently told that it does indeed always look like this.

---

[*] Obviously, it's Jocelyn.

My current office door is glass, but true to type it displays an out-of-date poster promoting a project long finished, a rainbow Pride sticker with the solar spectrum on it, and a brief description of what I do with my time, scribbled in marker pen for an open day in 2017, which explains that the incumbent is most likely trying to understand galaxies. My previous office had a better cartoon selection, including one from, I think, *The Far Side*, which shows a view of an observatory that is home to an enormous telescope projecting perilously out of the slit of its dome. Two astronomers, identified as scientists through the laws of cartoon physics by the white coats they wear,* peer through it to see a brick wall. Having reached the end of the Universe, they conclude it's time for us all to shut up shop, close the dome, and head home.

This idea of completing our reconnaissance of the Universe draws on an old tradition in astronomy, when the subject was less about understanding the forces and processes that produce the Universe around us and more about simply mapping the skies. Occasionally, the distinction between the two is described as astrophysics versus astronomy, though the terms blend together.** This older way of doing things is on display whenever you visit the great observatories of the past.

Take the Royal Observatory in Greenwich, for example, which sits on a hill outside Central London, high above the

---

* The last time I wore a white coat at work was for first-year undergraduate chemistry; my chemistry career was so brief that the thing never really developed the patina of stains that distinguishes the natural-born chemist from us mere dabblers.
** There is a practical definition. If I find myself wanting to talk to someone, for example on a plane, I say I'm an astronomer. If not, "physicist" usually does the trick and ensures silence.

Thames with fabulous views over the city. The main observatory is distinguished by a magnificent set of buildings designed, according to the original architect Christopher Wren (responsible for St. Paul's Cathedral, but also an accomplished astronomer), "for the observer's habitation, and a little for pomp." Most visitors to Greenwich today come to stand astride the line set into the observatory courtyard that marks the prime meridian, the line of zero degrees longitude acknowledged by mapmakers worldwide, though formally only after a nineteenth-century conference agreed how to standardize terrestrial coordinates.[*]

The prominence of the Greenwich meridian is a reminder that this observatory was established not to investigate the origins of the Universe and to understand how stars work, but simply to try to map the sky. An accurate star map was essential to navigation, and hence to the effectiveness of countries' navies. Though governments often squabbled over their own proprietary results, by the late nineteenth century the sheer scale of the task meant that things had had to become much more collaborative. At this point, a group of observatories, led by the one in Paris, embarked on a collective attempt to record the fainter stars in the sky, aiming to produce a significant improvement in the catalogs available. The most productive of the twenty or so contributing institutions throughout the life of this project, known as the Carte du Ciel, was the newly founded Vatican Observatory,

---

[*] If you stand on the line and ask your phone for your longitude you may get a surprise: when GPS was set up by the American military, they used an outdated model of the Earth's shape and cared most about getting things right in the continental US. The error a quarter of a world away in London adds up to a few hundred meters, creating a gap between the real and GPS meridians.

nestled in the hills above Rome. The observatory was set up because the Catholic Church was keen to establish the Vatican as an independent state and was busy creating all the trappings a modern nation would need: a central bank, a post office, and an observatory were all added in quick time.

The resulting Carte du Ciel catalog was based on more than eight million photographic plates, fragile objects covered in photosensitive emulsion, placed at the focus of a telescope and exposed to capture faint light from the sky. Together, those plates contained the positions of more than four million stars. Impressive though this total is, a lack of sustained funding and investment meant that it failed to match the ambitions of the project planners, who had intended their catalog, the first to take advantage of developments in photography, to go much deeper and see much farther than it eventually did. Sporadic efforts to improve continued long into the middle of the twentieth century, and this attempt to create a shared map of the sky has a long legacy. When I visited the Vatican Observatory a few years back, the Jesuit astronomers who work there used the Carte du Ciel project as an example of their ability to dedicate time to long-term, less glamorous observations rather than being distracted by chasing the latest scientific fad, hunting for exoplanets or looking for gamma-ray bursts.*

Sky-mapping has continued. Efforts from ground-based observatories have been surpassed by two missions from ESA, which takes great pride in leading the world in celestial

---

*   They also showed us their most precious meteorite, one that had been handled by three separate popes.

cartography. Orbiting telescopes Hipparcos and now Gaia have mapped the position and distance of more than a billion nearby stars, accounting for roughly 1 percent of the population of the Milky Way. Both missions used the observed parallax to calculate the distance to the stars that they were observing (see Chapter 6). The parallax is, as even the nearest stars are so far away, tiny, to the extent that the failure of nineteenth-century astronomers to detect it at all, around any star, was a major issue in debates about the size and structure of our galaxy. Gaia must work with incredible precision, taking into account the myriad other effects that might subtly affect the position of its telescopes. To give you an idea of the difficulty of the task, consider that although it is stationed one and a half million kilometers from Earth, at a gravitationally stable point known as L2, those processing its data still need to worry about which side of Gaia is illuminated by moonlight, the faint pressure of which might throw off its readings.

Mapmaking, whether on the Earth or in the sky, has often required such precision, but the result of all of this care and attention has been some marvelous surprises, many of which involve the dynamics of the galaxy. Astronomers can take advantage of the fact that by measuring the same stars again and again during the course of its multiyear mission, Gaia has been able to see how they move. We are able to use the results to watch the future of the galaxy, seeing the familiar shapes of the constellations dissolve as stars whizz about. Orion falls apart quickly, for example, while five of the seven stars that make up the most familiar northern asterism of all, the Big Dipper, stick together for many millions of years, revealed as not only a pattern that we see from Earth but also a true moving group of stars that formed, and which travel the Milky Way, together. Gaia has revealed that many stars belong to this sort of loose cluster, each one representing a burst of star formation that took place somewhere in the galaxy. We can even hope, one day soon, to be able to find, somewhere among the billion stars within our databases, the Sun's siblings, reminders of the time when our star, too, was forming as one among many in its natal cloud of gas and dust.

We can also turn the tape around, running the movie of this stellar dance backward to view our galaxy's history. This seems to be a form of galactic archaeology,* an attempt to understand the past from what survives in the present, and Gaia's results have been revolutionary. While the general assumption had been for many years that the Milky Way has lived a relatively quiet and calm life, at least by galactic standards, I wrote in Chapter 7 of the Gaia results that have shown us that the galaxy's disc contains the remnants of several systems that have collided with our galaxy, adding their bulk to our own.

Gaia has been studying individual stars too, showing that many more of them than we expected are roiled by starquakes, disturbances that originate deep in their cores and cause shock waves to ripple through their outer atmospheres. Something far within these stars is amiss, and understanding what it is will help

---

*   Not, by strict order of my friend and actual archaeologist Alice Gorman, to be confused with real space archaeology, the study of humans interacting with space. Alice and colleagues have looked, among other things, at how astronauts use the mementoes and trinkets taken up to the International Space Station.

us work out what makes them, and their quieter relatives, tick. While groups of mostly Belgian astronomers puzzle about the implications of this,* others are combing the Gaia catalogs to try to find the oldest stars in the galaxy, unsullied by the heavy elements produced by subsequent generations. The Sun is a member of the Universe's third generation. The material it, and the rest of the Solar System, and indeed you and I are made of has been part of two previous stars, but somewhere out there in the galaxy should be a smattering of stars formed from the nearly pristine remnants of what was produced in the Big Bang. Others hope to utilize Gaia in the years to come to try to find the tiny movements of stars pulled first this way and then that by the gravity of their orbiting planets, and still more astronomers are keen to use it to try to add to the census of asteroids in the Solar System.

The data flowing from Gaia, a marvel of early-twenty-first-century technology, is a near-perfect implementation of the vision to map the sky that drove generations of astronomers at Greenwich and other observatories across the world. In that the mission provides data that can be used by anyone in the world, it also represents a form of very modern survey science. When I started working seriously on data from galaxies beyond the Milky Way in 2006, instead of the trips to telescopes in Hawai'i I was used to as a PhD student,** I started working with data

---

* The world center of astroseismology is Leuven, a town otherwise famous for Europe's longest bar.

** It was sheer hell, as you can imagine, to have to visit that tropical paradise, one of the most unusual and interesting places on Earth. In mitigation, I should point out I spent most of my waking hours in the dark.

already safely captured and stored to disc by the Sloan Digital Sky Survey, which operated a telescope in New Mexico.

I often joke that Sloan is what you get when particle physicists build you a telescope. Deeply informed by the work of researchers from the US particle physics center Fermilab, who were interested in cosmology as part of the attempt to discover what the Universe is made of, and situated on a site that receives three hundred clear nights a year, Sloan's aim was to map the positions of a million galaxies and measure their distances. It was built as a tool with the single purpose of producing just such a catalog, rather than an observatory, a traditional astronomical facility that would do different things from night to night. Choosing between these two ways of operating is a trade-off that's common in everyday life—do I make a choice to do the one thing that will solve a particular problem, for example by getting pizza delivered for dinner, or do I put more effort into solving problems I haven't even thought about, for example by actually filling the fridge with vegetables for the week, even if it takes a little bit longer?

Massive and increasingly expensive science projects tend to be experiments, not observatories. The cost of an instrument that will do a specific thing, like Sloan's mission to produce the first good three-dimensional map of our local Universe, is easier to justify than something that might do many particular jobs but not concentrate on one. That this particular project achieved much more than was planned was due in large part to the presiding genius of a cosmologist and instrument builder named Jim Gunn. Jim is a small and softly spoken man. I once interviewed him on a summer's day, with Jim sitting cross-legged on the lawn, picking at the daisies as we discussed the problems of being scientists in a large Universe that seems resistant to helping us uncover its secrets. I was a little too in awe of Jim, and our topic a little too close to my own work, for it to be a very good interview, at least for the purposes of TV, but I tried to get him to explain what combination of his theoretical expertise—he was responsible for much of the theory of the pulsars discovered by Jocelyn Bell Burnell, described in Chapter 7, and for some of the foundations of modern cosmology—and his deep knowledge of how to build telescopes led him to insist on one crucial tweak to the design of Sloan.

At a meeting in the Hilton Hotel attached to Chicago O'Hare Airport, in one of the windowless conference rooms in which it specializes,* a group of astronomers gathered to hash out the details of the project. To do its principal task—mapping the location of a million galaxies—there was no need to take decent images of them, any more than when mapping the world you'd choose to represent London with an intricate drawing of the buildings clustering along the Thames and the arch of Wembley Stadium, when a pink dot would do nicely for the whole city. If the Sloan builders didn't need detailed images of each galaxy, they could build a worse and cheaper camera or, critically, spend less time looking at each system, speeding up the whole survey. Either way, money could be saved.

Several people at that meeting, though not as far as I know Jim himself, have distinct memories of him pushing hard for the survey not to take this easy route, but to record the best

---

*   I've attended conferences there and nearly gone mad as a result.

image of each galaxy that had ever been obtained. That is where he had started with the project, trying to work out how to make the most of the electronic cameras that are now found in every smartphone, but which in the 1980s were being developed specifically to replace those old-fashioned and clumsy photographic plates for use in astronomical work. The justification for doing such a good job of taking pictures of galaxies was vague, but no one wanted to extend their stay in the O'Hare bunker by arguing with Jim, and so that small change of plan was approved.

It made all the difference. Sloan was still a machine built for cosmology, but as it turned out the survey design, including those images, was perfect for understanding galaxies. What's more, it became astronomy's greatest engine for making serendipitous discoveries using its images in ways that its builders had not thought about at any point during its construction. Gravitational lenses, distant galaxies whose light was magnified and distorted by the gravity of nearer systems, were just the start. Sloan also found a whole bunch of nearby brown dwarf stars, and studies of the asteroids that crossed the field of view of its camera were among Jim's own personal scientific highlights—no mean feat for what was supposed to be a purely cosmological study. The galaxy pictures proved fascinating too, and I got involved in helping hundreds of thousands of people online sort through them via the Galaxy Zoo citizen science project.* Having that kind of attention paid to each image proved a very effective method for identifying the most unusual galaxies, and our

---

* See my book *The Crowd and the Cosmos*, available from all good booksellers.

volunteers were soon pointing us to systems that, though previously lost in a catalog, upon inspection proved to be worth a second or even a third look.

Time on the Hubble Space Telescope is, as we saw in Chapter 6, precious. The task of deciding what it should point at and when is made more complicated by the fact that it is, famously, in space, and more specifically in low Earth orbit, just over five hundred kilometers above our heads.

Placing the telescope in an orbit just above the atmosphere, rather than somewhere more distant and convenient for astronomy, was a deliberate decision, made in order to facilitate the task of servicing it with the space shuttle. The shuttle stuck pretty closely to low Earth orbit, and its flights to Hubble were among its most far-flung and risky missions. Over the years, that meant Hubble's cameras and instruments could be upgraded, but it came at a cost. When we plan observations from here on Earth we have to worry about whether the targets are rising or setting, and whether the Moon interferes,* but at least the telescopes stay still. Hubble is traveling at twenty-eight thousand kilometers an hour relative to us on the ground, completing an orbit every ninety-five minutes, and it is very important that it never points at the blindingly bright Earth,** making the choreography of scheduling an extremely difficult task.

Just occasionally, it doesn't make sense for the telescope to move to a new target straightaway if the Earth is in the way of where it should, all other things being equal, point next. For the

---

* The greatest single contribution to astronomical productivity we could make would be to get rid of the Moon. No one would miss it, I'm sure.
** Or the Moon. Get rid of the Moon.

last few years, unusual objects found by Galaxy Zoo volunteers searching through the Sloan data have been targeted during these gaps. On more than a hundred occasions, the world's most famous telescope has pointed at something first identified by someone without any formal astronomical training browsing an image on the web. Hubble points at each target for just eleven minutes, but that's often enough to create something marvelous. Our team meetings frequently start with the man in charge of these observations, Bill Keel from the University of Alabama, introducing the latest marvel fresh from the telescope in his wonderful southern accent. Being among the first to see in detail the fireworks of star formation induced in colliding galaxies, the oddness of elliptical galaxies that turn out to harbor spiral arms within themselves, secret until now, and the drama of systems with dramatic polar rings that loop right around their hosts is my favorite part of the week.

Bill is an old-school astronomer. Not just because he once turned up on a trip to an observatory with me with a backpack full of cassette tapes, full of data from the 1970s and 1980s he hoped to digitize, taking advantage of the presence on the mountain of the last remaining machine in the world that would read them, but because like me his astronomy is rooted in looking up at the stars. Though as expert a physicist as any, Bill fundamentally is someone who knows his way around the sky, and that's maybe why it was he who took one of my favorite images of all.

Unless you know what you're looking at, the shot doesn't show much. It was captured by a camera left pointing at the sky in the same direction with the shutter open for an hour or so, turning the stars into streaks of light as they moved. Cutting through the center of the image is a long, straight line, our view of the Hubble Space Telescope, which happened to be passing through the field of view. And at the time that Bill and his camera were looking up at Hubble, it was facing outward, observing the most famous of Galaxy Zoo's serendipitous discoveries, Hanny's Voorwerp.* Discovered by Hanny van Arkel, a Dutch schoolteacher, this is a gas cloud the size of a galaxy. In the Hubble image, where we represented the Voorwerp as green due to the wavelength of light emitted by oxygen in the object, it looks, depending on your particular cultural reference points, like an evil Kermit the Frog or a Romulan battleship, hanging in space about fifty thousand light years from a brighter spiral galaxy, IC 2497. It turns out that the Voorwerp is glowing because of activity around the black hole at the center of this neighboring galaxy. This black hole is quiet now, but it must, fifty thousand years ago, have been growing rapidly. Material launched then along a powerful jet driven by the black hole's twisting magnetic fields is causing the bright Voorwerp gas to glow now, making it a light echo, an object that tells us about the past.

Fifty thousand years ago, the activity in IC 2497 would have made its center shine, with the glow from material falling onto its black hole bright enough to be visible with binoculars. Sometime in the next fifty thousand years, the Voorwerp will disappear as the jet powering it switches off. This result, and studies of the very many similar objects found by eager Galaxy Zoo volunteers who wanted to follow in Hanny's footsteps, some of which have become targets for our Hubble snapshot

---

* "Voorwerp" means "object," or, colloquially, "thingy."

program, has taught us that, rather than being neatly sorted into those that are active and those that are not, black holes in the center of galaxies flip back and forth quickly between these two states. Indeed, the supermassive black hole at the center of the Milky Way, Sag A*, which we met in Chapter 7, may only recently have been in a vigorously active phase, though there's no sign of such drama now.

Running projects like Galaxy Zoo has let me get back to experiencing the joy of serendipity in my research, though nothing quite prepared me for the discovery made by volunteers on our Planet Hunters project. Just as Galaxy Zoo volunteers sorted through images from Sloan, the task we set visitors to the Planet Hunters website upon was to study data from NASA's Kepler space telescope, looking for the faint dips in brightness that occur when a planet passes in front of its parent star. Kepler was built for this job, and it did nothing but stare at the same patch of sky in the summer constellation of Cygnus for three years, recording the brightness of its stars with exquisite precision. Planets are small compared with their parent stars, and their expected effect is not large. A typical planet detected by Kepler might block less than 1 percent of the star's light; but even so, spotting such dips in brightness sounded, before we got going, like a relatively straightforward task.

As it turns out, it wasn't easy at all. The sky is full of what we call variable stars, whose brightness changes dramatically and often rapidly as a consequence of some internal instability. The Kepler team had been careful to include only quieter stars, without much known variability, in their sample, but these natural changes in brightness were still causing problems. The trouble is that if you look carefully enough, and record their brightness with sufficient precision, then all stars are variable. Many have starspots, just as the Sun has sunspots, colder patches that stick around on the surface for a few days or weeks. Because they are darker than their surroundings, they reduce the brightness of the stellar disc when they are visible, and, as we can't produce a resolved picture of the target star to see what is causing an episode of dimming, they look just like the transit of a planet in the data as they rotate across the face of the star. The Sun also pulses gently, and many stars rather less gently, the roiling of their turbulent surfaces often studied by astroseismologists hunting starquakes. Just as Earthly seismologists study earthquakes in part to understand the Earth's interior, so studies of the changing stellar brightness caused by these events can tell us about the inner workings of stars. They are useful, but the cumulative effect of all of this activity makes the data noisy and finding planets difficult.

That's why we asked Planet Hunters volunteers to look through the data. They've been responsible for finding all sorts of planetary systems, including—remarkably and unexpectedly—the only planet yet known to inhabit a four-star system. This consists of two tightly bound pairs of stars, spinning around each other, even while both pairs orbit the center of mass of the whole. Our planet revolves around not one but two of these stars, with a circular orbit that surrounds one of the pairs. This makes it a member of a small class of "circumbinary" worlds, but it also means that its sky, when two bright suns and two more distant, brilliant stars are all above the horizon, must be quite a sight.

The biggest surprise of all, though, was the behavior of an unprepossessing star recorded in the catalog as KIC 8462852, but which is now known as Boyajian's Star. Shortly after Kepler started watching this source, it dipped, just briefly, in brightness, before returning to its former brilliance. Interesting, perhaps, but a single dip doesn't really tell you anything. It could easily be a one-off, a giant starspot appearing momentarily, the appearance of an unusual stellar storm, or even something up with the camera. A few months later it happened again. Could this be a planet? If so, it would appear again after the same interval; we waited eagerly, but there would be no third dip.

The first two events couldn't, therefore, have been caused by a planet. Planets are there or they are not; they don't complete a couple of orbits and then stop for a rest, or decide to take it easy on the next loop around their star, or vanish suddenly into space. The mystery deepened a year later when the star dimmed once again, this time by about 20 percent, before returning to normal a few hours later. Stars don't do this! Specifically, no other star in the Kepler dataset—the other 199,999 studied by the satellite—ever showed this dramatic a change or put on such a show. A little more than a year later, things got even weirder, when over the course of a month and a half the star put on quite a display, fading and brightening rapidly with no conceivable pattern. Our volunteers named it the WTF star.*

---

* A name we managed to get into the scientific literature only by convincing the journal editor that WTF stands for "Where's the flux?" as we were searching for the missing light during periods when the star was fading.

What was going on? We started by checking the boring explanations. The camera wasn't malfunctioning, as no nearby stars showed signs of similar behavior, and it wasn't interference from any nearby star whose light could have washed over the camera and caused errors. The star in question seemed normal, stable, and middle-aged and, when not in one of its episodes of dramatic change, nothing of note. One of the volunteers responsible for drawing our attention to the star, Daryll LaCourse, thought the dips might have been caused by a dusty disc, similar to that from which the planets of the Solar System formed. The idea was that such a disc could eclipse the star at odd angles, resulting in the dramatic dips that we see. The size and duration of the dips would then be governed by the details of the disc's internal structure and orbital mechanics, and this would explain why each successive episode looked so different. It's not a bad idea, but we realized that if there was that much dust, enough to cause such significant dips, near a star, it should be glowing brightly in the infrared. However, when we looked there was nothing unusual going on in that part of the spectrum; the star was just as faint as a normal one of its type, unencumbered by dust, should be.

So what was happening? A small team of professional scientists and volunteers led by Tabby Boyajian (now at Louisiana State University) had assembled to bang our heads against this question, but we weren't coming up with anything fast. Pressed by the journal editor, who was reluctant to publish a paper that said we'd found a weird thing but not what it was, we eventually came up with a slightly half-baked idea. We blamed comets.

Comets are, as we saw earlier, mercurial things. Take Comet Biela, one of the first returning, periodic comets to be identified. It was discovered in 1772 and returned in 1805, and then on cue in 1826. At the time, several researchers calculated its orbit and made predictions leading, I suspect, to some satisfaction when it returned as expected in 1832, though on this particular trip to the inner Solar System it caused something of a fuss. The vagaries of orbital mechanics meant that on this return the comet would cross the Earth's orbit, which caused consternation in the press about the likely effect of cometary gases on life on our planet. The alarm didn't seem to be assuaged by the fact that the Earth wouldn't pass through that part of its orbit until nearly two months later, making a close encounter with anything left by the comet unlikely.

The brouhaha made the comet famous and its return was highly anticipated; it was indeed seen again in 1845. And then it vanished. Predicted returns in 1859 and 1865 produced nothing, but in 1872, the third return since the comet had been seen, the skies around the forecasted date of closest approach were illuminated not by a comet but by a spectacular meteor shower. Accounts of this storm suggest that as many as six thousand shooting stars an hour were seen, a stunning display that was repeated on the next return a few years later. That appears to have been the end of Biela's comet, though its remains are presumably still orbiting the Sun, if no longer intersecting the Earth's orbit.

We reckoned that if a comet had broken up around KIC 8462852, then its pieces—cometlets—would still be in orbit around the star. If they passed in front of it, as seen from our point of view, each would cause a dip in brightness. If we strung the pieces out along the orbit, each one could create a separate dip, which would then explain the full range of behavior we'd seen since we started monitoring the star. It seemed like a solid enough idea, not least because a string of comet bits could be placed around the star without needing to add much to its infra-red budget.

That was good enough to get the paper published, but there are a couple of flaws with this theory. One set of objections, which we quickly heard about from several people who actually knew about comets, was that any such object capable of block-ing this much light from the star would have to be massive, at least a thousand times bigger than any comet we'd ever seen in the Solar System. The odds of substantial pieces surviving after breakup for very long are low, so the largest comet ever discov-ered would have to have decided, by complete coincidence, to break up just as we started to observe the star. What's more, as no other star among the hundreds of thousands studied by Kepler showed this behavior, such an event would have to be really rare. Comets were looking unlikely as an explanation for Boyajian's Star.

A second objection to our comet theory was more subtle. The problem was that it could be used to explain almost any-thing the star did. Given the freedom to arrange my cometlets how I like, I can just move them around to match any dips that are observed in the star's brightness. It isn't that our theory is wrong, it's just that there's nothing the star can do to disprove it, as long as you're happy to add more comet bits or change the size of them to account for each subsequent dip.

That objection also applies to what quickly became the most popular, or at least the best-known, explanation for what Boyaji-an's Star was up to. I first became aware of the paper released by Jason Wright, of Penn State University, and his colleagues when I picked up the phone in the office on a quiet Monday morning. I am, to say the least, not the most agile of thought before the first three coffees have kicked in, and so I wasn't really in the best state to respond cogently when the voice of a reporter on the end of the line said: "I hear you've found aliens." Wright and company had suggested that our star's behavior could be explained by the presence, in orbit around it, of what they called an alien megastructure.

The idea is simple, and is usually called a Dyson sphere or swarm after the visionary physicist Frank Dyson, who popularized the notion that such objects might exist in the 1960s, though others later developed the concept. Any civilization worth its salt will need power, and any truly advanced aliens will probably have gotten to the point where they're obtaining it from orbiting solar panels. As these increase in number to meet the power needs of a growing highly technical civilization, they will envelop the star, blocking its light and producing the effect that we're seeing.

Such an idea is not completely crazy—I've been to a seminar at which a forward-thinking proponent of traveling the galaxy suggested that we as a species should get on with demolishing Mercury in its entirety, and using its material for solar panels as soon as possible—but it is vulnerable to the same criticisms as our comet idea.* We haven't a clue about the preferences intelligent aliens have for arranging their solar panels, whether they would prefer one large structure to many small ones, or whether they would leave gaps in the shield for practical or even artistic reasons, and so we can display the solar panels in any pattern we like. Just as with our comet idea, any sort of dimming can be explained by this theory without trouble.

Still, the idea that astronomers had suggested this might be aliens caused a stir. I'm especially fond of the national newspaper** that accused Jason's team and my collaborators of covering up evidence for aliens, even though we'd literally published papers detailing the thinking and the evidence on the internet. All the attention made us more determined than ever to solve the mystery of Boyajian's Star, and to try to prove that it wasn't really aliens. All we could do was to keep an eye on the star and hope it misbehaved again. Kepler had finished its primary mission, prevented from staring at its favorite patch of sky by a malfunctioning guidance wheel, but a network of small robotic telescopes that span the world was pressed into service to check in on Boyajian's Star periodically.

In the meantime, what we knew about the star, and the claims that had been made about it, had intrigued others, among them a specialist group of experts who spend their days worrying not about optics and electronic cameras but about the

---

*   Avi Loeb has also recently suggested that 'Oumuamua might be a fragment of a disintegrated Dyson sphere.

**   Not named for legal reasons, but you can guess.

photographic emulsions used for most of the twentieth century to gather precious light from the skies. A century ago, the advent of serious celestial photography made new discoveries possible, with the ability of a photographic plate to soak up light over many hours giving researchers the chance to capture and study objects that would otherwise have been much too faint.

Not that their lives were easy. In many of the giant reflecting telescopes that dominated twentieth-century astronomy, like those on California's Mount Wilson or Mount Palomar, light captured by the large primary mirror is reflected up to a focus point high above it. Here, in a small chamber known as the cage, astronomers would perch in the dark, carried from side to side by the telescope as it moved between objects, changing plates when necessary and hoping that their efforts would be rewarded in the darkroom. It was, I'm told, a specialized profession; tricks of the trade like tasting each side of the plate with the tongue to establish, in the pitch-dark of an observatory dome, which side had the emulsion and thus should be exposed to the sky, are now half-forgotten, but the plates, and the images of yesterday's sky that they contain, still exist, carefully filed in observatory and university libraries.

They have also been scanned, their data recorded and converted for a digital age. Interpreting what these records show is tricky, as the condition and behavior of plates taken in some cases decades apart can vary widely; but historical observations of Boyajian's Star seemed to show that it had been behaving very oddly indeed for some time, long before anyone had thought of launching a space telescope to look for exoplanets. In addition to the short-term changes the Planet Hunters had identified, it seems the star had been fading gradually over the course of the last century.

I think this was a blow for the idea of aliens. Obviously, one can imagine a sufficiently advanced civilization doing pretty much anything,* and I suppose you could say that this gradual dimming reflected progress as more orbiting solar panels were being added to the fleet; but, really, the idea that we could have spotted something as grand and spectacular as a Dyson sphere in the exact century in which aliens were working on its construction seemed a bit much.

In any case, we soon had our definitive answer. The star started dipping again, and this time Tabby and her team were ready with what used to be one of the most powerful tools at the astrophysicist's disposal: Twitter. Alerting the world that something was up, she galvanized astronomers to look again at the star. In telescope after telescope, camera after camera, and dataset after dataset, professionals and amateurs around the globe watched the star dim in brightness before returning to its normal self. But there was a catch. Setups that recorded red light showed a smaller dip than those that were sensitive to blue light; how much the star faded depended on the color.

This result was the key to understanding the star. It's simply not the way you'd expect a star being occulted by a solid object

---

* At a recent workshop on how to look for signs of alien civilization, I popped, just for a moment, to the loo, and returned to find that what had been a quite technical conversation about machine learning had become a philosophical chat about whether sufficiently advanced beings might make stars flash together or in sequence as a form of art.

to behave. Anything solid, including a fleet of orbiting alien solar panels, should block all colors of light equally. Instead, in each episode we had witnessed, it seems the star was being hidden from view by something like a cloud of dust, which would scatter blue light more than red. Dust could have been left over from a period of planet formation, created by a more recent event like a collision between two planets, or even produced by the star itself.

Dust is a good explanation for what we see, but the solution to this most surprising star's behavior only opened up new questions. Where did the dust come from, why was it found around this star and not the others in the Kepler dataset, and what will happen next? We don't have good answers to any of these, but my favorite theory, which also explains the long dimming found in the historical records, is a story about a planet that strayed too close to its star. Torn apart by gravity, such a planet could be reduced to rubble, which we now see blocking the star's light. Much of it could also have fallen into the star, and the theory seems to suggest—though we haven't tested it—that eating a planet would cause the star to brighten, temporarily, for a period of a few centuries. We may be witnessing the aftereffects of its substantial lunch, with the dust cloud that is responsible for its fading being nothing more than the leftover crumbs on the table.

Since the discovery a growing number of people have made it their business to keep an eye on stars behaving badly. Other extreme "dippers" have been found, and while we don't yet know whether we are witnessing a newly discovered phase of stellar evolution, an extreme version of the variability that many stars show in the course of their normal lives, or something else entirely, it's proving interesting.

Even well-known stars can behave in an unusual, if not so extreme, fashion. One of the most distinctive in the sky, Betelgeuse, in the constellation of Orion, is a familiar variable that in 2019 suffered an unprecedented fading.* There was great excitement. Betelgeuse is among the nearest stars that are massive enough one day to go supernova, and many felt that the slow fading might be a precursor of some more dramatic change.

A supernova just over five hundred light years away would be a spectacular event (and would please the German neutrino hunters I wrote about in the Introduction), most likely remaining visible in the daytime sky, even at noon, for months on end. As we haven't witnessed a supernova anywhere in the Milky Way since before the invention of the telescope, the opportunity for progress in understanding the violent death of stars would be immense. Yet I found that I couldn't join the ranks of those looking forward to Betelgeuse's demise.

Orion is easily recognized by identifying the three stars of his belt, located between Betelgeuse and the slightly brighter white star, Rigel, which represents the hunter's right foot. The constellation is a guide to the whole winter sky,** with the belt pointing up to Orion's prey, Taurus the bull, and down to Sirius, the dog star and the brightest in the sky. I must have looked up at it

---

* "Betelgeuse" is usually, loosely if irresistibly, translated from Arabic as "the armpit of the great one," and also, especially since the Tim Burton film whose main character was named after it, often pronounced "Beetlejuice."

** In the Northern Hemisphere; apologies to those in the south, who should turn this paragraph upside down.

thousands and thousands of times, and as a result I found the sight of the constellation's changed form profoundly unsettling. Though the stars do change and evolve, and the familiar patterns of the constellations will inevitably warp as they and we continue our orbits around the center of the galaxy, all of that happens on a timescale that is much longer than the lifetime of an astronomer.

To see something as familiar as Orion alter before my eyes was uncanny, for all the scientific potential Betelgeuse going bang would provide. Other astronomers, those who study massive stars, were too busy trying to work out what the cause of the dimming was to worry too much about how the skies looked. Betelgeuse is one of the few stars that are both big enough and close enough for us to see them with the most advanced telescopes as more than mere points of light. It normally shows up as a pale red disc, but images taken during its great fade with the Very Large Telescope showed that part of the disc was much fainter than normal; the dimming seemed to be affecting only half the star. A dust cloud, likely produced by the star itself, was obscuring our view.

If you're thinking of a star like Betelgeuse as a solid object, then I've given you the wrong impression of its nature. Our Sun is a ball of plasma, and rotates fast. As you descend rapidly into the star the density rises quickly, so though it is never solid, it is nonetheless a substantial thing. Inflated, puffed-up giants like Betelgeuse are much more tenuous, with their outer layers doing a decent impression of a good laboratory vacuum. Though it is true that placing Betelgeuse in the center of the Solar System would engulf the planets out to Jupiter, they'd be left traveling through a very diffuse cloud indeed, a red-hot vacuum. In this environment odd things can happen, including the slow condensing together of material to form dust particles that can block the light from the star. As we saw with Betelgeuse's recent misadventures, this dust can be expelled out into the cosmos. Such giant stars provide more than half of the dust that we see in the galaxy, in star-forming regions and around newly formed stars.

Indeed, much of the raw material from which the Solar System, and by extension the Earth and you and I, formed must have been produced in a similar giant, and now we were watching it happen in real time. Looking up at the familiar shape of Orion last night, with Betelgeuse recovered and shining away as brightly as ever, I now know that I'm looking at a factory that is producing raw material for future planets. Instead of thinking of its upcoming immolation as a supernova, I can imagine its contribution to worlds still to be born, out there in the vastness of the cosmos.

On a remote Chilean mountaintop, in the high, barren arid lands of the Atacama Desert, sandwiched between the Pacific coast and the Andes Mountains, a new window on the sky is about to open. The Vera C. Rubin Observatory is building a telescope as large as some of the biggest in the world, with a main mirror more than eight meters across, designed not for targeted observations but rather to survey the whole sky. Once operations start in 2025, this specialized instrument will scan the whole sky roughly every three nights, focusing light onto the world's largest

The beautiful structure of the Vera C. Rubin Observatory, on top of Cerro Pachón, Chile. Inside, the telescope that will conduct the LSST survey is nearly ready for action. *Credit: Rubin Obs/NSF/AURA*

digital camera, a 3,200-megapixel behemoth that is the size of a small car and which weighs about the same as half an elephant.

Such a camera is needed to take in as much of the sky in one go as possible, and it is an engineering marvel. I love the small touches that have gone into making it as perfect for its task as possible. For example, it has a shutter, just like any other camera, which is used to cover the chip between exposures. In this case, though, the camera is so large that having the shutter retract and then deploy after each exposure would take too much time. Objects on one side of the camera, nearest the shutter, would end up being seen for noticeably less time than those on the other side of the field. Instead, a special mechanism has been designed that means that the shutter always travels in the same direction, ensuring a uniform exposure.

The telescope and camera will be used to conduct what the observatory, in a slightly twee fashion, calls the Legacy Survey of Space and Time,* a name that at least reflects the survey's ambition. The LSST survey should make a big difference to all of astronomy.

It should, for example, find more supernovae in its first two months than we've observed in all of human history, thrilling both those trying to understand stellar evolution and cosmologists who rely on these distant explosions to measure the expansion of the Universe. It will provide the deepest images of large parts of the sky that we've ever seen, tracing in delicate filaments the streams and tails of tenuous material that record collisions and interactions between galaxies, and with more than ten million objects expected in its final catalog it will complete our first census of the Solar System, from the asteroid belt to the distant Kuiper Belt. It should even resolve a long-running controversy about whether there lurks, on the outer edge of the Solar System, an as-yet-undiscovered ninth large planet.

Such marvels do not come easily. The Vera Rubin Observatory is the product of decades of work by astronomers and engineers, software developers, and support staff, who collaborated first to make the case that it should be built, then to design and

---

*   The name comes from a previous name for the observatory, the Large Synoptic Survey Telescope, or LSST. When it turned out that no one knew what "Synoptic" meant, they changed the name but kept the acronym. I wanted to rename it the "Large and Slightly Stubby Telescope," but that lacks grandeur.

construct it, and to prepare for the data that it will send our way. The latter is an enormous challenge; a new fiber-optic cable has been laid between Santiago, Chile, and Florida, just to carry the data to the computers where it will be processed before being sent to astronomers around the world, who are expecting thirty terabytes of images from the camera each and every night, the same amount of data as contained in seven and a half million images of my dog taken with a decent mobile phone camera.*

These images of the sky can also be stacked together, so that over the course of the decade in which the telescope will operate we will slowly see deeper and deeper into the Universe; but the real challenge comes when we compare one night's observations with another, revealing on a scale never seen before how the sky has changed. The observatory's systems will identify objects that have changed brightness, or that have moved, from image to image, and send out an alert. It is hard to estimate precisely, as no telescope this big or instrument this powerful has even been used to do this sort of thing, but we're expecting something in excess of ten million separate alerts a night.

Hidden among them will be those supernovae and the trails of millions of asteroids, but also the faint flickering of galaxies as material falls into their central black holes, the signatures of exotic explosions like gamma-ray bursts, and the occasional blink of a star as a planet passes in front of it. We also need to face the fact that, if you have a big enough telescope, we will see that no star is stable, with pulsations and seismic activity driven by the fusion in its core, or the presence of starspots on the surface

---

* I seem to be approaching this rate. He is rather photogenic, though.

causing changes in brightness that the observatory's camera will pick up. Kepler showed us that understanding this behavior was difficult even with a nice uniform dataset, but under the gaze of this new instrument the stars will be seen to twinkle as never before.

There are too few astronomers in the world, and not enough time, for us to plan to look at this wealth of data ourselves. Plans are afoot to recruit volunteers via the Zooniverse citizen science platform that hosts both Galaxy Zoo and Planet Hunters, as well as to build powerful machine learning routines capable of sifting the data for us, catching a few nuggets of gold in the form of a distant supernova or nearby asteroid in a sieve and presenting them for inspection. Both approaches have merits,* but necessarily focus on searching for what we think we know will be there.

The Vera C. Rubin Observatory will have cost more than a billion dollars, by the time you take into account the expense of building the thing, analyzing the data, and running the observatory for the ten years of the main survey—money contributed by taxpayers in the US, the UK, France, Chile, and elsewhere. Though this is a lot of money, it will have been spent over several decades; even in the US, which is contributing most of the cost, it works out as much less than one cent per taxpayer per year. I am normally reluctant to talk about the cost of the satellites, probes, and instruments we use, believing that the benefits of discovery, of investing in inspiring science, outweigh the outlay. But a billion is a lot of money and, given the number of competing projects out there, not easily won.

---

* Have you read my other book?

The few hundred scientists who together made the case for spending such a large amount of money on this telescope did so by imagining the science it will do. There is a "science book," two hundred pages of description of everything that can be done with the data. Finding a few thousand supernovae will allow us to measure whether the mysterious speeding up of the expansion of the Universe is continuing at a constant rate or whether it is changing, giving us a clue to its nature. Identifying tens of thousands more near-Earth asteroids will allow us to reduce the threat to our planet from incoming rocks, and finding something between ten and a hundred 'Oumuamua-like objects will tell us whether our first interstellar visitor really was so unusual. Pinning more than ten billion stars, an order of magnitude more than Gaia can see, precisely to their positions on our maps will reveal the structure of the Milky Way, and on the grandest of all scales we will make the best map of the large-scale structure of the Universe, revealing its clusters and superclusters, its empty voids, and the cosmic web that envelops all we see.

Looking through the science book now, more than ten years after the first edition was written, I'm still excited about much of this work to come. Yet experience tells us that, before too long, the things we think we'll do with the new view of the Universe provided by the observatory will be eclipsed by the unexpected. The lesson from the stories in this book is that whenever we have looked longer, deeper, farther, or in new ways at the Universe, it has surprised us. From the advent of radio astronomy, which stumbled on a sky where distant galaxies shine more brightly than nearby stars, to the revelation of the microwave background by two men looking at radio interference, big breakthroughs seem to come from opening up new frontiers. Discoveries, from Jocelyn's pulsars to the sudden arrival of 'Oumuamua, also tell us that paying attention to the changing sky, looking for objects that alter in brightness or suddenly appear and then fade like Boyajian's Star, is a profitable way to encounter surprising new things. The LSST survey is designed just for this, and it would be staggering if it didn't find new transients, new types of dramatic explosion that can tell us about the Universe. There must be astronomical events we have never witnessed, just waiting for us to build a telescope large enough and a camera sensitive enough to see them.

The idea that this new frontier is just a year or so away is tantalizing. Thinking about it, I find myself staring once more at the sky, my eyes drawn to a faint fuzzy patch in the constellation of Andromeda. This is the Andromeda Galaxy, the Milky Way's nearest large neighbor, and at 2.2 million light years away it is the farthest thing that you can see with the naked eye. The light hitting my eyeballs tonight left Andromeda around the time that the first species of our genus, *Homo habilis*, appeared on the plains of eastern and southern Africa, and now here it is, providing me with a glimpse of our neighboring galaxy.

The realization that objects like Andromeda are completely separate from the Milky Way, and that each is a system of several hundred billion stars, came at the beginning of the twentieth century at the end of a series of arguments that lasted for decades, and it may have been the single discovery in the telescopic age that most changed our view of the Universe. From imagining a sky populated with objects that all belonged to our admittedly still pretty large galaxy, we were jolted into recognizing that, just as the Sun is nothing but an ordinary star and the Earth most likely simply another planet, our galaxy is merely another bunch of stars drifting through an impossibly and unimaginably vast cosmos. The sense of nearly overwhelming awe that I began this book by describing comes, I think, from this single realization, which puts the scale of the cosmos beyond our capacity to imagine.

A second change in our view of the sky is under way right now, with the discovery that the Earth is just one of many, many millions of planets in the Milky Way. When I started my career a couple of decades ago, it would have been possible to argue that planets were rare, and that the existence of the Solar System was just another cosmic coincidence, the result of drawing a winning ticket in a cosmic lottery. We now know that, instead, planets are incredibly common. The same physics and chemistry that worked to sculpt the familiar eight planets* operate throughout the galaxy, producing planetary systems of remarkable diversity and range. Almost every type of planet we can imagine, and everything included in even the most outrageous science fiction, has been found, from ones with twin suns in their sky to strange lava worlds whose orbits keep them perilously close to their parent stars. It remains to be seen how many of these systems and worlds actually look like our own, but it's still true that, looking up at the sky today, your imagination can skip from star to star, thinking about the worlds that might exist. Go outside again and look up. Knowing that most of the stars that I can see have planets transforms the familiar constellations from abstract

---

* Yes, and Pluto.

patterns to a map of places that we might, one day, visit. The galaxy is full of possibilities.

The Vera Rubin Observatory is, I believe, about to usher in a third transformation. Building on the work of a century, much of it described in this book, this new telescope will show us the night sky as a dynamic place. Though the view as I look up from my backyard will still be the same from night to night, changing only to reflect the slow turning of the seasons, I do feel my view of it will be irreversibly altered once LSST is up and running. If knowing that Betelgeuse might once again fade from its current reassuring brightness is already informing how I look at Orion, then how different will it feel when, with a few clicks on a laptop each morning I'll be able to call up yesterday's alerts and trace within them new asteroids, new supernovae, and the dramatic and not-so-dramatic modifications of millions of stars living out their lives?

Thinking of this new and dynamic Universe makes my head spin with possibilities, and the most exciting thing is that I have no idea what we are going to find. As the stories in this book make clear, the only sensible thing to do is to prepare to be surprised. Our attempts to understand the cosmos, to piece together the unlikely story that's led us to exist here, gazing back up and out at it, have taught us that the key skill in existing in this vast and wonderful cosmos of ours is being alive to the possibilities of serendipity. The Vera Rubin Observatory is designed to help us conduct the simplest and most rewarding of scientific pursuits: looking up at the sky. So let us do so together, and see what we can find.

ACKNOWLEDGMENTS

The first article I ever had published was about serendipitous discovery in astronomy. I have been thinking about how discoveries are made and how we can be surprised by the Universe, it turns out, for a long time. I thus probably shouldn't have been surprised that this book, on a similar theme, had a long gestation, as I tried to work out what stories I wanted to tell. Throughout, I relied on encouragement from the patient and wise Rebecca Carter, as well as the team at Janklow & Nesbit. My editor, Susanna Wadeson, has from the start been supportive and critical in perfectly judged measures, as well as patient when other demands on my time got in the way of promised writing. Thanks, too, to the team at Basic Books, who is responsible for this American edition; with their guidance I've made a very few changes from the original UK version. Much of the material gathered here was first encountered in my academic work in Oxford (where the White Horse provided a comfortable nook for writing on many Saturdays), or with the BBC's *Sky at Night* team (especially Chapter 5), and I should like to thank my patient and smart colleagues in both of those worlds. Other stories are those I've encountered through the wonderful staff and volunteers in the Zooniverse, who crop up every so often in these pages.

Chapters 3 and 6 also overlap slightly with articles I've written for the *London Review of Books*, which provides an occasional and welcome venue for my words. I'm also indebted to those colleagues and friends who have taken time to share their stories, or have had to listen to me trying to work out in real time what it was I was trying to say. Various audiences have had the same experience; I'm particularly grateful to Teresa Anderson and the team at Bluedot, the unique music and science festival that takes over Jodrell Bank Observatory for a weekend each summer, for letting me experiment on their stages, and to Robin Ince, Trent Burton, and everyone at Cosmic Shambles for the opportunity to talk to their people, from the Albert Hall to firepits in Northampton gardens. Brooke Simmons, Liz Dowthwaite, Emily Carrington Freeman, Jane Greaves, and Michele Dougherty proofread parts of the text and provided sage guidance. Mistakes, as ever, remain my own.

I started working on the book in the early stages of the COVID-19 pandemic, and much of it was written during those most uncertain of times. This is not, except in that coincidence of timing, a pandemic book, but I do hope that its ideas about looking up and being open to surprise are of some consolation to anyone going through challenging times. Gazing at the sky certainly has always helped me remain grounded on this planet, whatever else is happening.

GLOSSARY

ALMA   The Atacama Large Millimeter Array—a powerful telescope, sensitive to microwave wavelength radiation, positioned high in Chile's Atacama Desert and operated by the European Southern Observatory.

CMB   The cosmic microwave background: the faint glow of microwave radiation, received from all parts of the sky, that is attributed to the afterglow of the hot, dense Universe produced in the Big Bang.

dust   Small particles of mostly silicon or carbon, produced primarily in the atmospheres of large red giant stars. Each particle is, on average, about a tenth of the size of a grain of sand, and such material provides the raw material for the formation of planets. It is an old joke that any astronomical observation can be explained by evoking the effects of either dust or else magnetic fields.

ESA   European Space Agency, a collaboration of twenty-two member states, including the UK, responsible for missions including Gaia and Rosetta.

exoplanet   A planet around a star other than the Sun.

HDF   The Hubble Deep Field, the first deep image of distant galaxies obtained by the Hubble Space Telescope.

ISS   International Space Station, operated by American, Russian, European, Canadian, and Japanese space agencies, and continuously occupied since the year 2000.

JCMT   The James Clerk Maxwell Telescope on the Big Island of Hawai'i, designed to observe submillimeter—microwave—radiation from the cool parts of the cosmos. Originally built as a joint Dutch-British-Canadian project, it is now operated by a predominantly East Asian collaboration.

JWST   The James Webb Space Telescope, a large infrared space telescope that finally launched in December 2021 and which is now providing spectacular images and scientific results.

light year   The distance traveled by light in one year, and a standard measurement of astronomical distance. As light travels at three hundred thousand kilometers a second, this is nearly ten trillion kilometers. A lot, in other words.

LSST   The Legacy Survey of Space and Time, which will be carried out by the Vera Rubin Observatory, itself formerly known as the Large Synoptic Survey Telescope.

metals   To an astronomer, all elements other than hydrogen and helium are "metals." Hence metallicity, the proportion of a star or other object that is composed of such elements.

MPC   The Minor Planet Center, based in Boston, is responsible for cataloging the billions of small rocks and comets that throng the Solar System. They're busy people.

NEO   A near-Earth object—an asteroid or other small body whose orbit crosses that of the Earth and which might someday impact our planet.

regolith   The "soil" of the Moon or other body beyond the Earth; the distinction is that soil is produced and influenced by life. Regolith is just rock, though it is often made up of a combination of sizes of particles.

SETI   The Search for Extraterrestrial Intelligence, typically, though not exclusively, used to mean searches carried out with radio telescopes.

SKA   The Square Kilometre Array, an ambitious project to build sensitive arrays of radio telescopes across southern Africa and western Australia. The original design calls for a square kilometer of collecting area across the array, though the telescope will be useful long before that goal is achieved. Precursor projects, which include perhaps 1 percent of the final area, are already under way.

solar wind   The stream of often high-energy particles that flow out into the Solar System, away from the Sun. Trapped by a planet's magnetic field, these particles can collide with the upper atmosphere, exciting it into displays of aurorae.

STScI   The Space Telescope Science Institute in Baltimore, an independent research institution that runs

Hubble and JWST on behalf of the space agencies who fund them. Mission control for JWST is here too, so that the institute's staff can talk directly to the telescope.

UFO   An Unidentified Flying Object—usually presumed to be an alien craft, but technically anything in the sky that you can't identify. Because of the science-fiction stigma attached to the term, which perhaps now seems less serious, proponents of investigating such things now often call them UAP: unidentified aerial phenomena.

VLT   The Very Large Telescope, operated by the European Southern Observatory in Chile and actually consisting of four large optical telescopes, each with a mirror 8.4 meters across. They can work together or separately. A neighboring facility—the Extremely Large Telescope—is under construction, and will consist of a single telescope with a mirror 39 meters across.

XDF   The eXtreme Deep Field—a follow-up to the HDF.

FURTHER READING

The following is a combination of books and, where relevant, research papers that cover the topics in the text. Most of the papers are available online, and although some of them are technical, a lot can often be gained relatively easily by reading the introduction and conclusion and looking at the figures, leaving the details for later. (This is the technique I teach PhD students!)

I have listed papers using the common abbreviations used in the literature: *ApJ* (*Astrophysical Journal*); *ApJL* (*Astrophysical Journal Letters*); *MNRAS* (*Monthly Notices of the Royal Astronomical Society*);[*] *ARA&A* (*Annual Review of Astronomy and Astrophysics*); and *RNAAS* (*Research Notes of the American Astronomical Society*).

I have provided links to the arXiv preprint server, as the most convenient way of getting access to a paper. Where only an arXiv link is given, the paper in question was still in review at time of writing (July 2023).

---

[*] A publication that is not monthly, and which does not contain the notices of the RAS.

INTRODUCTION

E. Lakdawalla, *The Design and Engineering of Curiosity: How the Mars Rover Performs Its Job* (Springer Praxis Books, 2018). A detailed guide to the Curiosity rover and its marvelous cameras.

Quotation from *Last and First Men*, by Olaf Stapledon (Gateway, edition published 2021 in the SF Masterworks series).

CHAPTER 1: IS IT ALIENS?

History and background to SETI:

G. Cocconi and P. Morrison, "Searching for Interstellar Communications," *Nature* 184, no. 188 (1959), www.nature.com/articles/184844a0.

S. Scoles, *Making Contact: Jill Tarter and the Search for Extraterrestrial Intelligence* (Pegasus, 2017).

K. I. Kellermann, "The Search for Extraterrestrial Civilizations: A Scientific, Technical, Political, Social, and Cultural Adventure," in *Essays on Astronomical History and Heritage: A Tribute to Wayne Orchiston on His 80th Birthday*, eds. Steven Gullberg and Peter Robertson (Springer, 2023), https://arxiv.org/abs/2302.06446.

On the broader search for aliens:

S. Webb, *If the Universe Is Teeming with Aliens . . . WHERE IS EVERYBODY?: Seventy-Five Solutions to the Fermi Paradox and the Problem of Extraterrestrial Life* (Springer, 2015).

P. Davies, *The Goldilocks Enigma: Why Is the Universe Just Right for Life?* (Penguin, 2007).

CHAPTER 2: THE FOUNTAINS OF ENCELADUS

History of the Cassini mission, including its exploration of Enceladus:

M. Meltzer, *The Cassini-Huygens Visit to Saturn: An Historic Mission to the Ringed Planet* (Springer Praxis, 2015), (a).

M. Dougherty and L. Spilker, "Review of Saturn's Icy Moons Following the Cassini Mission," *Reports on Progress in Physics* 81, no. 6 (2018), https://iopscience.iop.org/article/10 .1088/1361-6633/aabdfb/meta (behind a paywall, sadly).

Since Cassini's discoveries, people have been suggesting missions to return to Enceladus. Most recently, there's TIGER:

E. Spiers et al., "Tiger: Concept Study for a New Frontiers Enceladus Habitability Mission," *Planetary Science Journal* 2, no. 195 (2021), https://iopscience.iop.org/article/10.3847/PSJ /ac19b7.

Recent observations with JWST show Enceladus is still very active:

G. Villanueva et al., "JWST Molecular Mapping and Characterization of Enceladus' Water Plume Feeding Its Torus," accepted by *Nature Astronomy*, 2023, https://arxiv.org/abs/2305.18678.

CHAPTER 3: THE SCOUT FROM REALLY, REALLY FAR AWAY

A good summary of what we know about our first interstellar visitor is given in:

The 'Oumuamua ISSI team, "The Natural History of 'Oumuamua," *Nature Astronomy* 3, no. 594 (2019), https://arxiv.org /abs/1907.01910.

And I also like the comparison with the Solar System's inhabitants given by:

P. Strøm et al., "Exocomets from a Solar System Perspective," *Publications of the Astronomical Society of the Pacific* 132, no. 1016 (2020), https://arxiv.org/abs/2007.09155.

The first suggestion that such objects might seed the formation of planets was here:

S. Pfalzner and M. Bannister, "A Hypothesis for the Rapid Formation of Planets," *ApJL* 874, no. 2 (2019): L34, https://arxiv.org/abs/1903.04451.

On this topic, it's hard to ignore Avi Loeb and his hypothesis that 'Oumuamua is an alien spacecraft, summarized in his book *Extraterrestrial: The First Sign of Intelligent Life Beyond Earth* (John Murray, 2021).

The most detailed response to the book is given by Jason Wright, Steven Desch, and Sean Raymond: https://medium.com/@astrowright/oumuamua-natural-or-artificial-f744b70f40d5.

CHAPTER 4: CELESTIAL VERMIN

An unusual take on asteroids (and how and why we should explore them):

M. Elvis, *Asteroids: How Love, Fear, and Greed Will Determine Our Future in Space* (Yale University Press, 2021).

There's a recent review of asteroid families:

B. Novakovic et al., "Asteroid Families: Properties, Recent Advances and Future Opportunities," to appear in CeMDA's topical collection on "Main Belt Dynamics," https://arxiv.org/abs/2205.06340.

The technical description of Winchcombe as a meteor:

S. McMullan et al., "The Winchcombe Fireball—That Lucky Survivor," accepted by *Meteoritics and Planetary Science*, https://arxiv.org/abs/2303.12126.

And the meteorite is described:

A. King et al., "The Winchcombe Meteorite, a Unique and Pristine Witness from the Outer Solar System," *Science Advances* 8, no. 46 (2022), https://www.science.org/doi/10.1126/sciadv.abq3925.

CHAPTER 5: PENGUINS OVER VENUS

The story of phosphine on Venus has been played out in a series of publications.

Jane's original paper:

J. Greaves et al., "Phosphine Gas in the Cloud Decks of Venus," *Nature Astronomy* 5, no. 655 (2021), https://arxiv.org/abs /2009.06593.

Followed by critical work:

G. Villanueva et al., "No Evidence of Phosphine in the Atmosphere of Venus from Independent Analyses," *Nature Astronomy* 5, no. 631 (2021), https://www.nature.com/articles /s41550-021-01422-z.

A. Akins et al., "Complications in the ALMA Detection of Phosphine at Venus," *ApJL* 907, no. 2 (2021): L27, https: //arxiv.org/abs/2101.09831.

And replies by Jane's team:

J. Greaves et al., "Low Levels of Sulphur Dioxide Contamination of Venusian Phosphine Spectra," *MNRAS* 514, no. 2 (2022): 2994, https://arxiv.org/abs/2108.08393.

J. Greaves et al., "Recovering Phosphine in Venus' Atmosphere from SOFIA Observations," 2023, https://arxiv.org/abs/2211.09852.

Speculative work on the life cycle of Venusian creatures is described in:

S. Seager et al., "The Venusian Lower Atmosphere Haze as a Depot for Desiccated Microbial Life: A Proposed Life Cycle for Persistence of the Venusian Aerial Biosphere," *Astrobiology* 21, no. 10 (2021): 1206.

CHAPTER 6: STARING INTO SPACE

Popular articles about the deep fields and their impact:

Nadia Drake in *National Geographic* (www.nationalgeographic.com/science/article/when-hubble-stared-at-nothing-for-100-hours) and Fabio Pauccini in *Scientific American* (www.scientificamerican.com/article/how-taking-pictures-of-nothing-changed-astronomy1/).

A nice review of science with the original Hubble Deep Fields:

H. Ferguson, M. Dickinson, and R. Williams, "The Hubble Deep Fields," *ARA&A* 38, no. 667 (2000), https://arxiv.org/abs/astro-ph/0004319.

A book-length review, with more of the history:

D. Nardo, *Hubble Deep Field: How a Photo Revolutionized Our Understanding of the Universe* (Compass Point Books, 2017).

Territory covered at a more technical level:

R. Williams, *Hubble Deep Field and the Distant Universe* (IoP Publishing, 2017), available at https://iopscience.iop.org/book/mono/978-0-7503-1756-6.

For a taste of the distant Universe, consider flying through the Ultra Deep Field:

https://svs.gsfc.nasa.gov/30687.

CHAPTER 7: LISTENING TO THE UNIVERSE

I'm indebted to G. Verschuur, *The Invisible Universe: The Story of Radio Astronomy*, 3rd ed. (Springer, 2015), for much of the background in this chapter.

K. I. Kellermann, "Grote Reber (1911–2002): A Radio Astronomy Pioneer," in *The New Astronomy: Opening the Electromagnetic Window and Expanding Our View of Planet Earth: A Meeting to Honor Woody Sullivan on his 60th Birthday*, ed. W. Orchiston, vol. 334, Astrophysics and Space Science Library (Springer, 2005), https://link.springer.com/chapter/10.1007/1-4020-3724-4_4.

See also W. Sullivan, ed., *The Early Years of Radio Astronomy: Reflections Fifty Years After Jansky's Discovery* (Cambridge University Press, 2010).

For a British view of the pioneering days of radio astronomy:

Bernard Lovell's autobiography is readable and informative: B. Lovell, *Astronomer by Chance* (Basic Books, 1990).

CHAPTER 8: THE OLDEST LIGHT OF ALL

The original discovery of the CMB:

A. Penzias and R. Wilson, "A Measurement of Excess Antenna Temperature at 4080 Mc/s," *ApJ* 142, no. 419 (1965), https://adsabs.harvard.edu/full/1965ApJ...142..419P.

The explanation by Dicke's group, published in the same issue as the discovery paper:

R. Dicke et al., "Cosmic Black-Body Radiation," *ApJ* 142, no. 414 (1965), https://adsabs.harvard.edu/full/1965ApJ...142..414D.

The description of the cosmology of the early Universe given here is now out-of-date in some aspects, but still unparalleled in reach and clarity:

S. Weinberg, *The First Three Minutes: A Modern View of the Origin of the Universe*, 2nd ed. (Basic Books, 1993).

I would also recommend:

P. Coles, *Cosmology: A Very Short Introduction* (Oxford University Press, 2001).

For those looking for something more technical, there's a good review here:

R. Durrer, "The Cosmic Microwave Background: The History of Its Experimental Investigation and Its Significance for Cosmology," *Classical and Quantum Gravity* 32, no. 12 (2015), id. 124007, available at arxiv.org/abs/1506.01907.

CHAPTER 9: THE CHANGING SKY

C. Lintott, *The Crowd and the Cosmos* (Oxford University Press, 2019), tells the story of the Zooniverse citizen science projects, including the discovery of Boyajian's Star.

The original paper:

T. Boyajian et al., "Planet Hunters X. KIC 8462852—Where's the Flux?," *MNRAS* 457, no. 3988 (2016), https://arxiv.org /abs/1509.03622.

And the suggestion that it was due to a megastructure:

J. Wright et al., "The Search for Extraterrestrial Civilizations with Large Energy Supplies. IV. The Signatures and

Information Content of Transiting Megastructures," *ApJ* 816, no. 1 (2015), https://arxiv.org/abs/1510.04606.

Jason also wrote a review of the many different ideas that had been suggested here:

J. Wright, "A Reassessment of Families of Solutions to the Puzzle of Boyajian's Star," *RNAAS* 2, no. 1 (2018): 16, https://arxiv .org/abs/1809.00693.

I'm still a big fan of the idea that the star has consumed a planet:

B. Metzger, K. Shen, and N. Stone, "Secular Dimming of KIC 8462852 Following Its Consumption of a Planet," *MNRAS* 468, no. 4399 (2016), https://arxiv.org/abs/1612.07332.

INDEX

Chris Lintott is a professor of astrophysics at the University of Oxford and the winner of the American Astronomical Society's prestigious Beatrice M. Tinsley Prize. He was previously at the Adler Planetarium, Chicago, and is best known as copresenter of the BBC's long-running *Sky at Night* program. He is the author of *The Crowd and the Cosmos* and coauthor of *Bang!!: The Complete History of the Universe*. He lives in Oxford, UK.